summer of gravel
and steel

summer of gravel and steel

a thru-hike of Alaska, 20 years after the first

Ned Rozell

CONTENTS

1. A notion — 5
2. First steps — 15
3. Thompson Pass — 30
4. Bears — 49
5. The Big Lonely — 63
6. Spark plugs — 74
7. A lot like me — 92
8. Birds — 99
9. Into the boreal forest — 107
10. Walking home — 120
11. Cora — 133

CONTENTS

12 | The real deal 144

13 | Longest day 158

14 | Isom Creek 173

15 | Yukon River 201

16 | Two quotes 210

17 | The tent 224

18 | Northern cat 231

19 | Trail angels 245

20 | Girls 259

21 | Exit the boreal 279

22 | Into the Anthropocene 284

23 | Classic rock 300

24 | The end 313

25 | Afterword 328

26 | About the author 336

To Tony and May and Brian Jackson and Drew Harrington and David Bartecchi and Clutch and my friend Fluffy. Tempus fugit, yes.

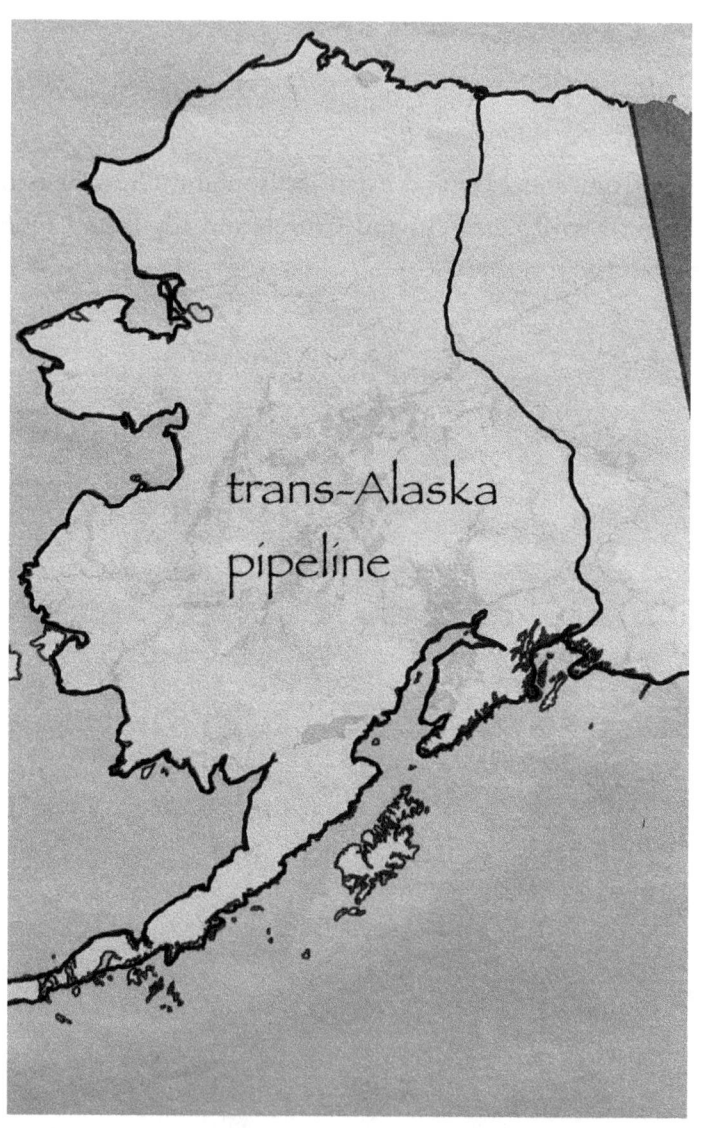

1

A notion

When I was in my mid-thirties, 20 years seemed like a long time. It doesn't seem so long now. Twenty years is a third of my life. Where did those 10 million minutes go since I walked across Alaska with my dog, Jane?

A few years after that four-month journey along the Alaska pipeline, my handsome chocolate Lab died at age 13. I somehow thought her brown head would forever be nudging my leg.

I was adrift when she left me. That somewhat prepared me, but not really, for the death of my father a few months after.

Mortality was not something I had pondered in my mid-thirties. I did not have enough experience with the end of life, regarding any species. However, if I had paid close enough attention, I might have heard a twig snap.

The new year of 2017 arrived cold and dark in Fairbanks. Fifty-one below at the airport, with ghostly ice fog that inspired nostalgia.

For the first time in the written record of Fairbanks, which extends back to 1906, the winter before (2015-2016) had not fallen colder than minus 30. I hadn't bothered plastic-wrapping our double-pane windows (triple is the way to go), and they didn't fog up from the cold.

In January 2017, the return of deep, hissing cold was a surprise, like seeing a friend walking the street years after last seeing him, only now with a shock of white hair.

A new year, lots of inside time. Time to think. Assess.

Fortunate, healthy, 54-year-old me had a notion to re-hike the same path across Alaska I had 20 years earlier. Why? Mostly because I could, with no body parts that hurt. Walking across the landscape has always been appealing for the human creature, hardwired to explore. And, for me, things seemed to be strangely falling into place.

On one ordinary day that winter, I typed Pipeline Hike 2 on my iPad calendar. Outlined in pink: the months of May, June, July and August.

A few years earlier, the University of Alaska Fairbanks

chancellor had put the notion in my head. I was meeting with him to see if he would fund me to write a glaciologist's biography. During small talk, before I asked for the money, the chancellor suggested I might re-walk the pipeline route on the 20th anniversary of my first hike. Not as some official university event, just because it would be an interesting thing to do.

I had not thought to repeat the experience before he mentioned it. One summer of walking that line, and the fun months of research and writing the book that came after it, seemed sufficient.

But it turned out to be one of those unexpected bits of conversation that keeps poking at you. It hung there, waiting for distractions to fade. The image of me walking across Alaska came to me night after night, as I lay in bed.

Much had changed in my life between 1997 and 2017, but there was one similarity. After a brief period without canine companionship, I again shared a house with a good dog.

In the prime of life at 3 years old, Cora would be a contrast to Jane's last good steps on that same trail at age 10.

Ten also happened to be the age of a being I hoped would accompany me for many miles on the pipeline trail: my daughter, Anna. That she was Jane's age during the last hike was coincidence, but it felt like serendipity.

And then there was the administrator for Alyeska Pipeline Service Company, who had the power to grant or deny permission requests to use the pipeline pad, a gravel road that runs most of the 800 miles next to the pipe. When I checked, the head of the right-of-way division happened to be the same man I had talked to 20 years earlier. That seemed unlikely, and somehow meaningful.

Another unusual thing about that summer: For the previous decade, my wife Kristen had been on trips all over Alaska, to watch geese and count oystercatchers and net warblers. In 2017, her schedule was blank.

For the previous summers of Kristen's fieldwork, sunny seasons with no nights, I was a stay-at-home dad. I pulled Anna in the bike trailer to the university's homey day-care as I worked half days and wrote newspaper columns. After biking home, I poured hot water into dried macaroni and cheese, and read to her at naptime before tucking her in.

When Anna snoozed, I looked at Facebook posts of friends on their packrafting trips all over the Alaska map. I felt like a chump, the forsaken one. Who signed me up for this? Oh yeah, me.

At the same time, I savored Anna's head falling into my shoulder as I read to her on the couch, and all the other cozy connections to this little human who shared my facial structure, my scent, my taste for caribou liver pate.

Things had changed since those days of around-the-clock attentiveness. Anna was no longer a 3-year-old. She could poop without adult assistance, and walk at a similar speed to adults carrying heavy packs. She could tell us when she was cold and needed her raincoat.

Also, at the time, a dip in worldwide oil prices had slowed a good deal of noisy activity in Alaska: BP and ExxonMobil cut back on their Alaska operations, resulting in fewer contracts for biological research companies, like the one for which Kristen works. She had no field time on the calendar. It would come, it always did, but there was nothing yet in ink. It was time for me to make a plan.

I wrote a one-page proposal to the pipeline right-of-way leader, describing me walking most of the 800 miles with my daughter, who would be 10 — just like my dog was 10 the first time. The symmetry! She would be out of school, and therefore in need of something to do instead of costly craft and rock-climbing camps. The efficiency!

It was an idealized plan, without much realism. The girl is like me — protective of her time and ideas, and stubborn. If a notion is hers, she's all for it. To buy more time to think, she greets proposals with a no, sometimes a maybe. In the preceding months, every time I mentioned the walk across

Alaska, I heard no. I was receiving a dose of myself in 10-year old female form.

But with every no, I tried to remember the way she moves when outside. While walking, her steps are buoyant. She skips. Her rhythmic skate skiing, practiced since she was five, mists me up as it nears perfection. During parent-teacher conferences, her fourth-grade teacher said she was a different kid in running club, smiling and chatty. A contrast to the quiet, furrowed brow she wore at her desk adding fractions.

If I just got her out there, I figured, she would enjoy doing what her mother and I love. Being outside, searching for soft, flat ground to pitch the tent, seeing what's around the next mountain. Moving, that conduit for joy.

I had no idea of what images might stick to a little human mind, but I remembered a few Velcro moments of my own. Our parents driving five kids in a Volkswagen bus from upstate New York to Maine, where we camped at Hermit Island. Every Fourth of July's celebration at the Bertrands' camp in old New York farm country, framed by rock walls piled during the Revolutionary War. I loved the musty smell of the big canvas tent when my dad zipped it shut. The delicious warmth of the Coleman sleeping bag when my skinny, sunburned legs brushed the plaid interior. The astounding volume of crickets. I assumed my two brothers and two sisters were as smitten. But as we fledged into adulthood I turned out to be most enchanted with what our Irish grandmother called "instant poverty."

A few weeks before I was to start walking, Anna accompanied me to the Fairbanks security desk of Alyeska Pipeline Service Company. On that April day we visited the Fairbanks Alyeska office, my blondie squinted as a security guard typeset information off my driver's license.

He was issuing me a "right of way use guideline." He called it a RUG. The RUG allowed me linear travel by foot from pipeline mile 0, at Prudhoe Bay, to pipeline mile 800, in Valdez. I would walk it from south to north, beginning near the deepwater port of Valdez, as I had 20 years ago.

The Fairbanks guard asked for other hikers' names. I told him Anna and Kristen. He typed them in and handed me a printed copy of the RUG.

Anna observed, but said nothing until we were outside, walking around piles of dirt-specked, melting snow. She strode on long, thin legs. A colt.

"Does that paper mean I have to hike the pipeline?" she asked.

"No, it just means you *can* hike," I said.

"Good, because I'm not doing the whole thing. This is your idea, not mine."

My heart sank. There was my typical reaction again, coming from a small girl who looked like me.

Waiting to see us in the back seat was little Cora, part Lab but taking mostly after her dad, a blue heeler. She wormed and wagged as Anna opened the rear door.

I folded my permission slip around my driver's license and tucked it into my wallet. I climbed into the driver's seat. Looking into the rear-view mirror, I knew at least one creature back there would soon be out of its mind with the freedom that paper would deliver.

I started to drive and thought a bit. Regarding the 10-year-old I helped create, I knew this: I would not force her out on the trail when she objected. We have done that. It drains the fun from girl, from wife, from dog, from me.

The girl's objection was a red light, a stop sign for the trip. Or was it?

If I could not get Anna out the whole hike, which was starting to becoming clear, we would figure it out. Or at least Kristen would. She enjoys complicated logistical problems. And Kristen saw the summer hike as something from which our girl might benefit.

With that little piece of paper — the RUG — in my pocket, we drove toward the west side of Fairbanks, and home. On the way, I saw a familiar, hunched figure, biking on the shoulder of the road.

Everett Wenrick worked as a salesman at Beaver Sports, which is probably where he was biking from. White hair spilled from beneath his scarred helmet.

Though his boating days were near done, due to failing eyesight and hearing, Ev had among many other adventures dipped his paddle in most every waterway in middle Alaska. Even the slow, boring ones, like the muddy Tanana River that arcs past Fairbanks.

As we passed Ev, I thought of his catch-phrases. One of my favorites was "Ski ya later," which he scribbled in cabin logbooks.

As he became a speck in the rear-view mirror, I remembered another Ev-ism, this one regarding pre-trip uncertainty.

"It's always right to go."

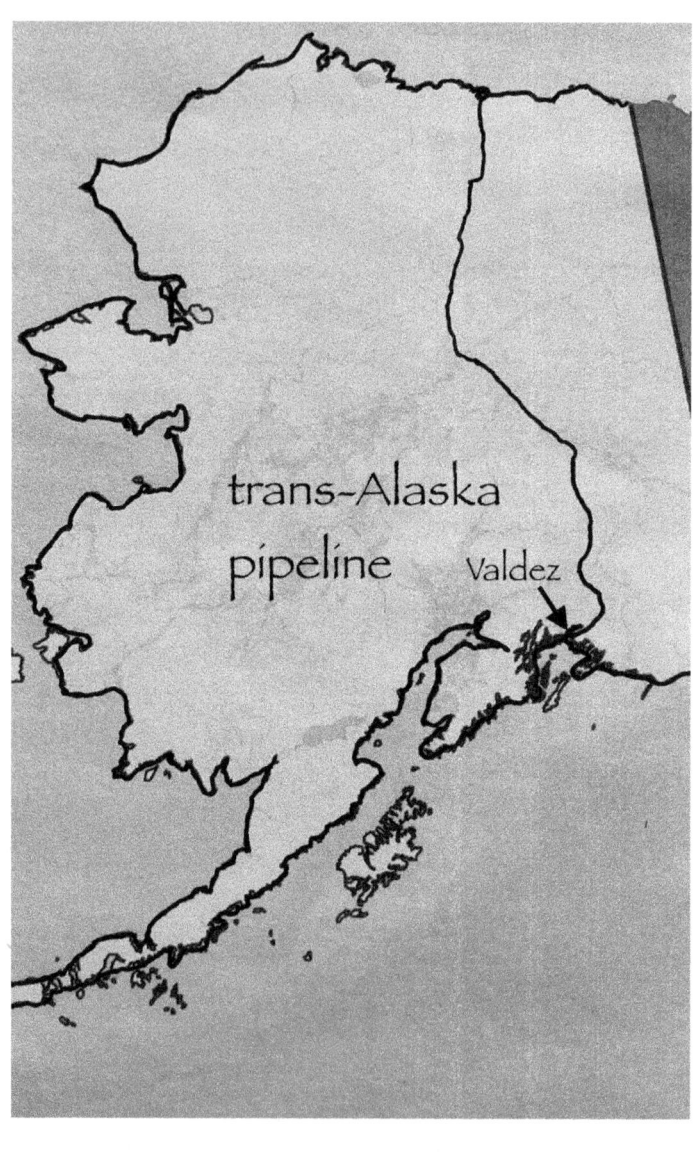

2

First steps

VALDEZ — On that last day of April, snow remained. Born of coastal winter storms that dropped two more wet feet with each atmospheric disturbance, the snow was a sad remnant of its clean, former self. Compacted by gravity, each layer smushed by more snow and, as the days warmed, further shrunken by rain. The snow became a drab, uniform coating on the sculpted landscape, a dirty white blanket.

Early that morning, before I started Pipeline Hike II, I visited Alaska adventurer Luc Mehl at his girlfriend Sarah's house in Valdez. I knew Luc from times we had both participated in the Alaska Mountain Wilderness Classic Ski Race, a point-to-point race through the backcountry: choose-your-own-route and carry as little as you dare.

Twice, we briefly intersected on that white field of semi-competition. He and his partners had won each of those races across 160 miles of snow and mountainscape. Luc finished enough days before me that he was sleeping at home by the time I arrived at the final checkpoint cabin. When I later watched his trip videos on YouTube, I could not believe how quickly he had moved over the same landscape.

I have teased Luc for messing up the Classic for me. He and his disciples cover wilderness ground so fast that the race organizer tacked 40 more miles onto ensuing routes. That change accelerated my aging out of the race. Several times, I had run out of all the food I could carry. I am not getting faster.

From my home in Fairbanks, I had messaged Luc to see if he wanted to join me on the mountainous portion of the pipeline's route above Keystone Canyon, 20 miles from Valdez. He did not say he wanted to accompany me, but invited me over for a cup of coffee. With Kristen, Anna and our dog Cora, I drove to the address he gave me.

Luc was taller than I remembered, lean and taut as steel cable. He had just completed a ski traverse in the Valley of Ten Thousand Smokes in Katmai National Park and Preserve. He and his partner skied up snow-capped volcanoes while carrying packs that weighed 80 pounds. Then they peeled climbing skins from metal-edged skis, locked in their heels, and skied down the mountains.

My proposed trip along the pipeline was of a different nature: No navigational challenges, a gravel road to follow.

I was trying to gather a bit of local knowledge of this place of soaring mountains and yawning valleys. I had not visited Valdez in years.

Luc told me of a friend's cabin I might use in a pinch. It was about a week up the highway/pipeline route. We sipped strong coffee with Kristen (Sarah was not home) as heavy rain wetted the snow outside the window. We said nothing about the weather, which was the kind of winter/spring transition day you might choose to curl under a blanket with a good book. Luc did not seem tempted to join me on my first days of the hike. I did not ask him.

After hugs with Luc, Kristen and I walked back to Anna, who had waited shyly in the car. We had driven down to Valdez from Fairbanks the day before.

I backed out of the wet driveway and drove again through Valdez, stealing looks across the water toward the marine terminal, where supertankers awaited a bellyful of crude oil at mile 800 of the trans-Alaska pipeline. Those behemoths would soon start a journey southward, to oil refineries in Oregon and California. I would soon start a journey northward, in an attempt to cross the whole of Alaska.

It was time to start the hike.

I then felt the pangs, wondering why I wanted to repeat an experience from 20 years earlier.

There, looping the long road around Port Valdez, they came to me: my parents and spirit friends who were a great part of my motivation to walk across Alaska once again.

On my mind-screen was tall, bent Dave Johnston, who froze — and later had amputated — a few toes on Denali during the first winter ascent of Alaska's highest peak. He had since climbed the high point of every state during winter. Dave once told me I should never do anything twice. At the time, I nodded and agreed. It's a big world. Lots to see and do.

But somehow I had decided to do this again. I had come this far. What kept me rolling toward Allison Point was a desire to have an atypical summer after a series of over-predictable ones. To step off the trail of normalcy I had walked for so long — father, husband, science writer — and live outside for a summer. And because Dave Johnston had recently suffered a stroke that left him unable to climb any hill.

Brian Jackson's big smile was also urging me on. During a recent autumn, Brian was in Wisconsin, handing a deer rifle to his sister. She was sitting up in a tree stand. When she grabbed the stock, the gun fired. Brian died.

He was a flamethrower of fun, superman-diving toward second base at the softball field or doing backflips at a Zoolander theme party. Brian was a decade younger than me. His death again reminded me of what a privilege it was to be 1) above ground and 2) the owner of a healthy, moving body.

As I drove, I could no longer call my parents to describe the mountains, the rotten salt smell of the tidal flats, the tingle in my guts. In 1997, I had written a letter, asking them to come along and walk a gentle section with me and Jane. My dad was then 65, Mom 63. Tony Rozell wrote back: *My old ass just isn't up for it.*

Mom didn't respond. Though I didn't know it, she had by then lost her beautiful ability to write.

Tony Rozell died three years later, of a bad liver, brought on by a diet of scotch and Planter's peanuts. By that time, Mary Rozell was losing her mind to early-onset Alzheimer's. My dad, who was perhaps physically capable of sharing a few steps, did not want to leave my mom at the time I wrote him the letter. We five Rozell kids had no idea how much mom had slipped by then.

A few years after that, I returned to my Dad's hospital bed just before he died, but not soon enough to hear him say goodbye. I regretted not buying a ticket home earlier.

I cared for my mother for a month after Dad died. In her prime, Mary Helen Rozell was a tiger on the tennis court. She was lovely, smart, strong. She was my doubles partner. She was my mommy, into whom I snuggled while we watched

TV, who hugged me and soothed me when a girlfriend dumped me.

In the end, as I stayed with her alone in my boyhood home, she was squeezing toothpaste on her comb. She was in her mid-60s when I placed her right hand in that of a friendly woman at the nursing home. Because I didn't think my heart could take it if she again pleaded with me to take her home, I hid in the bushes. From gaps in the leaves, I watched as that kind soul escorted Mom to her new room. My hiding was maybe the same action my mother had taken when she dropped me at the bus stop on my first day of school.

* * *

I had questions about myself. Was I, at 54, too much a creature of comfort to sleep on the ground every night? Would I find the walking exhilarating, or tiring? How much had that 800 miles of Alaska changed in 20 years? Would I remember what was around the next bend? Would I call Kristen for a ride home after a week?

When I became emancipated at 18, my observant mother said she never thought I would live a typical life, nor keep the same job for long. I was too much a nonconformist, who saw absurdity in the 9-to-5 model. I told her as much, when I knew everything there was to know at age 15.

Mary Rozell, who died in 2005, would be surprised today

to know that 20 years later, I had the same job I did then, as a science writer for the Geophysical Institute at the University of Alaska Fairbanks. I have stayed on for a few reasons: No one has ever told me to punch a clock, nor ordered me to write about a subject. In choosing those topics, I have travelled this giant peninsula from Attu to Ambler, from Southeast to Shungnak.

My previous hike along the route had provided me a good template with what to do regarding my job. Twenty years ago, I had walked into her office to ask my boss for the summer off. I was ready to quit, too, if that's what it took to gain the freedom to walk. She surprised me by countering with a proposal that I write columns from the trail.

In 2017, I asked a different supervisor, two removed from my first, that I be paid half of my salary for delivering newspaper columns from the trail. Though she was not excited about the plan, she respected it, and agreed.

I told her the hike would give me something to write about, no small thing when you've written one 640-word column each Thursday for almost half of your life. That's about 700,000 words that have appeared on the Sunday pages of Alaska newspapers.

Those hiking pieces, I hoped, would be a mental break from my typical science writing, and might attract a different type of reader. I imagined someone reading the end note and thinking the Geophysical Institute was a cool place. A public-relations soft-sell, which might attract free-thinking graduate students to study there. Maybe not, but I was excited for some first-person narration and some free-flow prose that wasn't checked by a scientist (though they have saved me from an oil-tanker load of errors over the years).

* * *

Unknown to me, my wife had, as she always does when I'm on a trip without her, hidden a note in my red backpack.

"I can't believe you'll be gone all summer!" it began.

"We will miss you and think about you and hopefully have daily check-ins. Anna and I look forward to meeting up with you soon!!"

In the months before, I had slowly convinced Kristen after I slowly convinced myself I was doing the second hike. I slipped in references during dinners and rolling time in the car. Yes, I would be sleeping on the ground all summer, and not in our back yard.

"Hey, you married the pipeline hiker," a friend from Jersey City told her when she discussed my summer plan. I laughed, and was relieved when he said that. Hell yes, I am the pipeline hiker, I thought. Getting out on this type of adventure is something I love, movement my spirit needs.

Kristen knew that, and I think she saw value in what I was doing, enough that she was on board. She wanted to hike the whole 800 miles with me and Cora and Anna, but she would take what she could get with her work schedule and Anna's school. That would amount to two week-plus stints of family hiking along the way.

Things became more real for both of us a few weeks before the start. I purchased, with her blessing, a satellite tracker/GPS/texter that fits in a pants pocket. That would allow me to check in with her every morning and night.

Kristen's fieldwork trickled in, but it remained light. Her sister Sarah was coming up from Colorado with her two pre-teen boys, Garrett and Brody, to explore Alaska (some of it with me). Sarah would help with 10-year old Anna.

* * *

On our drive to the start of the hike in Valdez, we picked up our hiking partners for the first few days — our neighbors Chris and Ian Carlson, and the Carlson's Labradoodle Freya. They jammed into our car with their loaded backpacks and curly wet dog, and we left their car near a bridge to which they would hike. Chris, Ian and I planned to reach it in three days.

Tires hissing on the wet road, we rolled to a stop at our

starting point, an access road that led maintenance trucks to the pipeline pad.

From left, Ned, Ian and Chris

Pausing on the slog up the pipeline-access road, I listened to the tires of our Honda as it zipped away from Allison Point. That car, obscured from view by needled branches of giant Sitka spruces, carried Kristen, who had to return to work, and Anna, who still had a few weeks of fourth grade to complete. They would spend another night in Valdez before returning to Fairbanks.

Their departure was similar to watching one's float plane lift off from a lake in the wilderness: the engine's whine fades to silence, and you wonder if you've packed enough food. Sure, we would be within a few miles of a major highway on most of this walk, but in spots we would be hours from any outside help. And the mosquitoes had already started to bookmark us.

As was to become common, I had a flashback, my first. This time to Takeoff Day 1997.

Then, I had also walked through snow, that time one mile west along a chain-link fence outside the Valdez Marine Terminal. The terminal is the end of the pipeline, where tankers receive North Slope oil, gravity fed from tanks on a hillside.

On my first hike, I wanted to start my walk as close as possible to pipeline mile 800, the coincidental round number. My dog Jane and I had postholed through unpacked snow to get to the fence that surrounds the Valdez Marine Terminal.

Now, I was taking an easier way from the highway to the pipeline pad, via a road constructed for workers and security guards in vehicles.

On the present trip, from the access road Chris, Ian and I intersected the pipeline pad at mile 799. We then turned left, toward Prudhoe Bay.

Thirty-four-year-old Ned would have found the one-mile shortcut unacceptable. But Ned with Silver Hair wanted to get rid of any thoughts of purity right away.

Instead of insisting of walking the entire pipe length, I would take opportunities that arose, like accepting offers to sleep inside instead of camping out every night. And, as I had learned from the previous pipeline hike, there are several areas of private property to avoid along the route, which can be skirted by walking the highway.

But that night I would be camping out with Chris and Ian.

Chris Carlson, my friend and neighbor, was born in Alaska while I was a teenager in New York. He grew up on a barley farm in Delta Junction. He is a lefty who bashes softballs into the right-field gap and reverts to a kid rounding first. That inner boy greatly resembles Ian, Chris's 12-year-old son.

Ian is my partner in the joyful blasting of Wiffleballs over his house during Home Run Derby on summer afternoons. Ian's mother Audra, teaching violin inside, is tolerant of the balls that fall short, banging the sheet metal roof and rolling back down to be caught for an out.

Freya, a sweet Labradoodle I call Fluffy, has sheep-like white fur, with which Cora loves to floss her teeth. If best-friend status exists between dogs, they have it, chasing each other blindly, often into the knees of any human standing around. They continued this fun even with Cora wearing her backpack. It was blue, consisting of two saddlebag compartments, each filled with an equal amount of dogfood in Ziplocs, as well as her collapsible plastic bowl.

Early on, the pair of trail revelers took out Ian, tackling him at the knees as they ran side-by-side at each other's necks. They blew him up. Ian, 85 pounds and slender, rose to his feet, saying he was OK, as Chris scooped snow from the hood of his son's jacket.

While we walked through the soggy mashed-potato snow amid misty rainforest spruce, a thought emerged: Would we find a dry tent site? The weather, somewhere between rain and snow, spattered our rain gear.

"No one likes a wintry mix," Kristen had said on the drive to Allison Point as she had read the forecast on her phone. Chris and I repeated that phrase to each other several times as we walked.

But we were out of the barn. Kristen, Anna, and our car were already miles away.

A Nome adventurer refers to escaping "the glue of town" in his online trip reports. On that last day of April at 3 p.m., plunging through saturated snow, a wintry mix darkening our jackets, the Carlsons and I had pulled free.

With those first steps in my new hiking boots came delicious relief: If it's not in my pack, I don't need it.

It was done. All the planning and decisions and food-packing and supply-stashing and permission-gathering and the little and big confrontations. I exhaled deep, and pulled in a lungful of clean mountain air, scented of spruce and pink salmon carcasses and wet moss.

After just three miles of walking, we found two grassy sites elevated just enough to be free of standing water. That would do for the tents. On any trip, I like takeoff day to be a short one, so as not to get too gassed. And to properly celebrate getting unstuck.

While Ian climbed a tall, gray, dead spruce, Chris and I pitched Chris's Clip Flashlight and my Quarter Dome 3 nose to nose. The dogs roamed free as we worked. They returned to us with no black bears following. We had seen no tracks in the snow in this ideal bear habitat of sharp mountains with fireweed shoots popping up on warm slopes. Bears might not yet have emerged from hibernation; another sign winter had not yet released its grip.

That evening, Night One of my trek, we were bundled in all our clothes for dinner. Chris, Ian, and I sat on the python roots of a Sitka spruce. That tree was sucking up northern rainfall when George Washington was crossing the Delaware River.

Our cookstoves propped on last year's leaves and needles, we boiled water from a nearby mountain stream. Eating our first dehydrated dinner, we smelled salt water and listened to the thrum of diesel engines. The boats floated a few miles away on Port Valdez, one of the farthest north harbors in America where the surface of the ocean does not freeze in winter.

In the dim forest, chestnut-backed chickadees flitted about. A female, eye streaked white like a warrior princess, perched on a spruce branch as thick as my arm. She was so close I could have brushed her feathery chest with my fingers. One second later, a male chickadee stuck a landing beside her. Wasting no time, he dabbed his tiny body atop hers, implanting the first seeds of summer. Game on.

3

Thompson Pass

The trans-Alaska pipeline snakes up and over three mountain ranges, a silver thread that stretches north to south across the state. Except where it worms underground, as is the case from Valdez to about 60 miles north.

Engineers would have buried the entire 800 miles of pipe if not for permafrost, frozen ground that is a relic of frigid times long ago, when cold hovering in the air penetrated deep beneath the grasslands and peat bogs. Because the four-foot diameter pipe was to carry oil warm as bathwater, planners routed it above ground in areas of frozen soil for 420 of its 800 miles across Alaska.

Less than 20 miles from Valdez rises treeless, snowy Thompson Pass, a low point through the Chugach Range. The passage is 3,000 feet above the briny waters of the Gulf of Alaska. The Chugach Mountains block moist, warm maritime air from middle Alaska. Denied the moderating effect of the immense body of the ocean, Interior Alaska is both warmer in summer and much colder than the coast in winter.

At the base of the pass came a couple of major route decisions, which the resilient snowpack made easy. Instead of the pipeline hiker, I would become the highway hiker.

As much as I wanted to walk the entire pipeline pad (the gravel road, adjacent to the pipe, whether buried or not), I thought it better to avoid some of the troublesome sections as the pipe exited Valdez. Too steep, too much of a slog.

Hiking it in those early spring conditions was like this: Set your foot, apply weight, sink into 14 inches of wet snow. Repeat. It was OK going uphill with a firm toe point, but downhilling with a flimsy heel-plant was a slip and slide.

I had sought purity at a cost 20 years ago, Then, from the floodplain of the Lowe River, Jane and I had huffed up a green ski slope of a hill, the pipe buried beneath, to where the line traverses above Keystone Canyon. That section must have been a headache for the route-planners in the 1970s. Instead of following the lead of highway-builders — who chose a low path through the canyon just above the glacial river — the pipeline engineers routed the pipe straight up a mountain until reaching a contour that paralleled the river.

The path to the roof of Keystone Canyon was about

the pitch of a stepladder. Engineers had designed the trans-Alaska pipeline to carry brown, viscous goop that didn't care much about steepness as long as jet engines within the pump stations were shoving it. I cared, though.

Twenty years before, Jane and I had found ourselves a few hundred feet above the paralleling highway, watching the toy cars and trucks below. The snow had not yet melted up there on the mountainside. The iced-over creeks that crossed the path made me wish for crampons. I carried none. While descending the glazed switchback access road north of Keystone Canyon, my feet had squirted out from under me. I lost grip on my shotgun. It skidded away, plastic and steel clicking downhill on ice and snow, not stopping until a bush caught the gun, barrel pointed at my head.

This time, I had the benefit of experience.

"I'm not going up there," I told Chris as we camped for our third and final night together, with the pipe's Keystone Canyon ascent looming above us. "With all this snow, even the south side would be scary. I'll find another way."

The obvious choice was to follow the Richardson Highway's route through the canyon, a few feet above the gray Lowe River. Chris, Ian and Fluffy would soon roll onto that pavement, to begin their seven-hour drive to Fairbanks.

As Chris and Ian drove away that morning, I felt both sad and relieved, alone for the first time. Twenty years ago, I had cried when my girlfriend departed from near the same spot.

Both times, the air was moist and heavy, but it was not raining.

This time I did not cry, though I was just as lonely with my friends gone. Maybe, just like the little girl who finally realizes that her parents will return after leaving her with the babysitter, I knew the solitude was temporary. I had learned to savor it.

Cora and I began walking on the asphalt shoulder of the Rich. There, we initiated a ritual we would repeat many times after being flushed to the pavement by less-than-hikeable sections. I grabbed her leash from my backpack hip belt, let gravity spill it to its three-foot length, and gave her a command.

"Come."

With her head warming my knees, I fastened the leash to a handy ring on the rear of her backpack. I threaded the nylon handle loop through my pack's hip belt before clicking it closed.

"OK."

Cora pulled the leash tight and maintained its full length, straight as a yardstick. She possesses a quality Jane did not have — she likes to lead.

A leashed walk with Cora in town is unpleasant, for the stress it puts on your wrist. With her fastened at my waist, she yanks, gently but constantly, helping me walk faster. The constant tension on the leash also made it easy to ferry her over to the side of the road when a truck approached.

We entered high-walled Keystone Canyon on the Valdez-bound shoulder of the Richardson Highway. I walked. Cora tugged. A few cars whizzed past, the drivers glancing left at us, as if suddenly seeing a moose and its orange calf in the willows.

Before long, we veered left, to a paved oval pullout for Bridal Veil Falls. There, drivers twisting through mountain walls stop and take cellphone photos of clear water vaulting off a mountain. I looked toward the hill and discovered for the first time a subtle path heading into Sitka spruce. It was the beginning of the Goat Trail.

Now a hiking path, the Goat Trail parallels the highway yet runs 100 feet above it. I had read about it before, in a history book on Alaska's Gold Rush and the All-American Route to the Yukon goldfields starting from Valdez. The trail's elevation was still a few hundred feet below the pipeline's Keystone

Canyon path Jane and I had crept along on the other side of the river.

The Goat Trail cut up into the forest, leading Cora and me above the highway. The entrance was steep enough that I had to kick my toes into the black soil while pulling myself up with stems of cow parsnip.

According to a note in a trailhead logbook, volunteers from Valdez had the autumn before snipped overhanging branches from the trail. They had left behind a fragrant tunnel of sap-oozing alders. Soon, Cora and I ascended to a point where the sound from cars disappeared. Snow endured on the trail, but in intermittent piles that were easy to step around.

When a stream of clear, lovely meltwater gurgled across the trail, I stopped. I dropped my pack to the greenery and felt the coolness of sweat evaporating from my back. I then dipped my pot in the stream and boiled water for tea and Ramen. It was time to celebrate being alone with Cora in the woods.

Sitting there with Cora curled up at my boots, me warm in my heavy sweatshirt that was a panic purchase at the Prospector in Valdez, I yelled.

"YES!"

I was elated to have found such a pleasant walkway adjacent to the highway, one I had passed in my car many times without realizing it was there. And, alone at that moment, drinking from a stream with my dog at my side, I realized again that the planning was over. I was walking across Alaska once again. Pure freedom.

Once wide enough to accommodate horse-drawn wagons, the overgrown Goat Trail continued along mountain walls, where a marmot whistled, perking Cora's ears.

The path stopped dead at an alpine creek, which had eaten it away. I could see the trail continued on the other side, past a few gravel braids. Flowing from a snowfield high in the mountains, the creek was one of the first I needed to de-boot for. There were dozens more in the next few months.

Strapped on my pack was a pair of Rics, a knock-off version of Crocs, bulbous foam sandals that seemed light as a handkerchief. Carrying my boots in my hand by their tied-together laces, I splashed through ice water that cleaned my feet, carrying specks of grit down to the Lowe River. Cora jumped right in, wetting her pack but finding enough purchase to trot right across.

The 12-mile Goat Trail had the character of the Pacific Crest or the Appalachian trails, except Cora and I were, as far as I knew, the only ones on it that day. That thought — of such extreme space — sent a pulse of happiness through my brain. It reminded me why Alaska, with its average of one person per square mile, has been such a good fit for me during the last 30 years.

After the trail had transited Keystone Canyon, Cora and I punched through drifted snow and twisted back on a descent to the Richardson Highway. Our few hours of wonderful road avoidance were over, but the Goat Trail, a nice surprise for me, had allowed us to get away from the pavement shoulder and blatting Alaska 4-by-4s for most of a day.

* * *

We crossed the highway, and regained the buried pipeline's path that paralleled it. The pipe's approach to Thompson Pass loomed ahead. I looked up and saw a straight white line resembling a ski jump, only 10 times as long. My belly tingled.

During pipeline construction 40 years ago, the pair of welders who joined the sections of pipe on that wall became known up and down the line. Wearing climbing harnesses and helmets, they had mated the pipe on the near-vertical.

I there had to make a decision. Crawl up that pitch like an ant, or hike around it on the highway. I had tried climbing the wall 20 years earlier. when the snowpack was less dense. On the third and steepest slant, I started avalanching rocks on Jane with each of my kicked-in toeholds. Did I really want to go through that again? Cora was a little mountain goat, seven years younger than Jane, but was it too much for her?

I mulled that question while camping for a night in spruce woods by Sheep Creek. The next morning, still undecided, I walked a portion of the pipeline pad that is also a subdivision road.

As I hiked, I distracted myself by wondering which was the home of Mike Maze, whom I had interviewed and wrote about in *Walking my dog, Jane*. I met him by chance back then, as he had recognized me when he was driving his truck home, and pulled over to chat. He said I should try to hike up the pipe's path up Thompson Pass; he had done it a few times.

Back then, he said the pipeline affected him in a way he didn't appreciate. Oil tumbling down from atop Thompson Pass in a freefall caused booms in the pipe there at the bottom. His house shook sometimes. Though I may have been imagining it in the silence, I thought I could sense a disturbance while walking, a vibration in my ears that thrubbed like far-off explosions. Twenty years ago, on the first hike, I had detected nothing.

I did not remember which driveway was Mike's. I was too shy to poke down any to look for him; I had not seen him since walking away from his house after interviewing him 20 years before. And, though spontaneous encounters are fun, my first instinct is to avoid them. I tend to steer away from confrontations, even good ones, unless I really push myself.

I continued past all the gravel driveways, breathing a sigh of relief when no barking dogs raced out. I saw the Richardson Highway crossing the buried pipe's path ahead. Beyond that, I saw the pipeline's straight-ahead cut up the pass.

Hell no, I was not climbing that. I resigned myself to some head-down travel on asphalt.

To traverse Thompson Pass (named for Pennsylvania congressman Frank Thomson, and misspelled by a USGS mapmaker), the Richardson Highway zigs and then zags. The long, lazy switchback added six miles to our day when compared to the straight-up, direct route of the pipe.

As I stepped over a guardrail and lifted Cora over like a suitcase (using a convenient handle on her pack), we were standing on the Richardson Highway.

Once again, thoughts of impurity crept in. Man, was I a loser for again not being able to climb straight up Thompson Pass. But that notion evaporated quickly, maybe because of the 20 additional years of mileage on my chassis. I chuckled to myself at the thought of a friend who stood next to me in a cold rain a few years ago, just before the cannon went off to start us up the hill in the Equinox Marathon.

"Nobody really cares how fast you're going to finish, except you," he said. "If you don't give a shit, it's a lot more fun."

I found that thought to be liberating, an epiphany. Though I'm competitive, there was relief in realizing I could race without needing to beat my rivals. A few times, after they crossed the line before me, they greeted me with a cup of Gatorade and a smile. I had made their day, because they finished ahead of me.

It wasn't always like this. The year I got together with Kristen, she and I staged a competition between us, knocking off personal bests in every running race in Fairbanks. She took about half of them, me the other. We were strong and fast and held nothing back. I even pulled a hamstring in the last few yards of a 5K, while surging to sneak up on another runner I had never beaten.

I was peaking, in my late 30s, and was set to run my best Equinox Marathon ever. Then, my father took his last breaths on his bed in upstate New York. I flew there to say goodbye, and to care for my mother afterward.

I came back to Alaska a changed person. One of the things that seemed to have disappeared was my competitiveness. I still wanted to do well in races, but I stopped searching for the results in the paper the next day.

* * *

Leashed again to my waist belt, Cora pulled as if she was

SUMMER OF GRAVEL AND STEEL

born to the task of yanking me up the hill. Pure, transparent water flowed just off the shoulder of the highway. Ditch water from Heaven. The melt from smothering snowfalls (50 feet piles up in Thompson Pass each winter) coaxed me to the shoulder for a drink. I treated the transparent water with my Steripen, even though no giardia cysts were probably tumbling within it. Cora and I sat down and shared some beef jerky.

Walking the highway was beginning to feel like a great choice. There was almost no traffic. Maybe ten times per hour, Cora and I veered to the side of the road to allow a southbound driver to pass. Most of the time, she did not need to be leashed, though I kept her locked in. While walking, I took in the chrome mountains and incised valleys we were slowly climbing above, something I always want to do while driving.

The smooth asphalt made for a quick turnover. The aerobic fast-walking was almost like going for a run, an activity I would not engage in that summer. As the sweat cooled me, I recognized the endorphins, and welcomed the feeling of pure happiness.

In good time, Cora and I zagged westward and reached the pass, marked by a green-and-white highway sign as being 2,678 feet above the salt water I had tasted on my fingertips a few days before. On our march up the gradual slope of major

Alaska highway, we had left behind the bitter scent of Sitka spruce. The air smelled clean and icy.

The crowns of alpine tundra plants poked through the snow, reaching for spring sunshine while their stems were chilled and smothered, but the pass surrounding the highway was still a white world. Near the Thompson Pass highway sign, a blueish snowdrift rose 15 feet higher than my head. Plow drivers had cut a sharp trench through all that snow to maintain Valdez's road connection to the rest of the world.

A few hundred steps beyond the pass, Cora and I found a gravel airstrip that Department of Transportation drivers had plowed. I pitched our tent on gravel off the main runway. There, we listened to a noise we hadn't heard for months: The spooky trill of Wilson's snipes, birds fresh up from the tropics to feast on the insects here.

Alaska's main appeal for billions of returning songbirds is the explosion of delectable bugs in every corner of the state, and at every elevation, including high snowfields onto which spiders and ants are blown.

Like me, those snipes, little balls with pencil beaks hovering above, might have felt a few days early. The great insect hatch had not yet commenced at Thompson Pass. I could have left the tent mesh unzipped, but we sealed things up to cut the breeze and gain a few degrees of warmth.

When I looked up from the sleeping bag, Cora curled warmly against my legs, the blue ice of Worthington Glacier filled the tent window. Perfect.

The next night, after another day of highway walking, we slept on dry soil under a bridge, in order not to pitch the tent on wet snow. It was our sixth night on the trail, and most of the miles we walked had been on pavement.

Spring would catch up someday soon, but we had pushed the calendar back a couple of days by climbing into the high country. Still, no matter how snow-covered the landscape had looked while whizzing past in a car with Kristen and Anna days earlier, there was always a dry, tent-size rectangle of ground somewhere.

After dusting off and packing the tent from under the bridge, Cora and I again began another day of hoofing it up the highway shoulder.

On the seventh afternoon, I saw something I had not seen since Chris and Ian left: people who were not zipping by in cars or trucks.

As Cora and I walked downhill on one side of the Richardson Highway, a young boy jogged toward us on the other side of the road. Following him at a relaxed pace was a woman. His mother.

"Crush that uphill!" I yelled.

The boy kept running, ignoring me. The woman smiled at me and trotted behind him.

We were surrounded by rocky mountains and clear, fast-running streams. That stretch, from Valdez to Glennallen, is one of the least-populated sections of paved highway in Alaska. No homes, no gas stations, no wires, just the roar of water and poplar trees bending in the breeze. Where had those two come from?

Soon, the boy and the woman passed us again, running back the way they had come.

"Do you want coffee?" the woman asked.

"Yes."

"Come over to our house and have a cup," she said. "First driveway on the right, in the aspen trees."

There was a subdivision in that gorgeous wilderness? I never knew. Cora and I turned right on the subtle gravel road.

L.J., home alone with five-year-old Logan, was a nurse, her husband an anesthesiologist.

He was that day working in Bethel, a few hundred miles

away in southwest Alaska. L.J. sometimes left to do the same thing. The work enables them to live there at "46-Mile," one of many hidden mini-communities along Alaska's highways.

In their yard, Logan was a perpetual motion machine throwing sticks to another, Cora. Normally, he occupied a good deal of his mother's time, including homeschooling. That day's science experiment had Logan placing two plastic cups of water in the freezer, one laced with salt. His assignment was to ponder why one froze and the other did not.

As I watched L.J. coaxing Logan through his hypothesis, I recognized the drill. I thought of a truism from another dad as we sat on metal bleachers, watching our kids at a playground: The minutes drag, but the years fly.

I dropped my pack and sat on a log in front of their outdoor fireplace. L.J. handed me a plastic mug with black coffee she had brewed inside.

Handsome and confident in the presence of a stranger, L.J. struck me as a self-sufficient, independent Alaskan. Dressed in nylon running pants, a t-shirt, and with a circular digital watch as thick as her wrist, she was a different breed from the white-bearded Alaska sourdough. She was perhaps the new version, searching for the same adventure in this new frontier of paved roads, supermarkets with starfruit from Costa Rica, and airports offering flights to anywhere.

Logan and L.J.

I also thought: Here it begins. Just like 20 years ago, I'm starting to meet people on my trek, to learn their stories, hear their unique perspectives, see where they live.

L.J.'s home was centered on an aspen plateau above a beaver pond. She had brewed coffee in the shadow of nearby Mt. Billy Mitchell and its magnificent hanging glacier, a spaceship of ice that crashed into a mountain.

Bored with stick-throwing, Logan hid a toy truck for me to find. He placed it somewhere by their house, an immense tin-sided log structure that is under construction, like many of the places at 46-Mile.

To live in those quiet, raw mountains, L.J. and her husband sleep away from home for long stretches, working where they are needed, banking cash. Sometimes, the whole family will tack plywood over the lower windows and go live in Fairbanks for a month while working at the hospital there. I had perhaps passed L.J. and Logan as we pushed our carts while shopping at Fred Meyer's.

When I finished my coffee, I went searching their tidy acreage. I found Logan's truck under a five-gallon bucket. It was then time to go. Mother and son hiked me out their road.

On the subdivision's road of gray rocks, we walked past other 46-Mile homes, each invisible from the nearby highway.

At one, neighbors were helping each other insulate an outhouse. That heatable privy would someday belong to Lindsay, a newborn nested in a sling around the torso of Lindsay's mom, who wore a wool sweater that smelled of wood smoke.

Because Lindsay's cute, wrinkled forehead was so close, I leaned over and smooched it.

"Welcome to the world."

Lindsay's bearded dad, sawdust on his sweatshirt, smiled and shook my hand.

His wife asked me what I carried for bear protection.

"Twenty years ago, I carried a shotgun 800 miles. This time, just pepper spray, which I didn't carry last time."

I lowered my hand to quick-draw position, mocking a gunfighter's pose. The pepper-spray cannister waited in the thigh pocket of my nylon cargo pants. Lindsay's dad nodded in recognition.

Lindsay's mom then said words that came back to me on my nights in the tent with Cora. In rubber boots and torn canvas pants, Lindsay's mom looked like a person who was comfortable sleeping outside.

"You don't need a gun," she said. "You have a dog. That's way better."

4

Bears

Without fail, when I met someone while I was hiking, bears crashed every conversation.

Have you seen any? What are you doing to protect yourself? What will you do when you see one?

Then came the bear story. Everyone I met had one. Usually more than one. Each tale revealed the teller's fear and reverence for a brute whose habitat remains over most of Alaska. A noble, athletic creature we have not pushed out.

Even as bruises of humanity swelled over its map, Alaska continued to host healthy populations of bears almost everywhere in the state, except the Aleutian Islands. The ones near the coast, which feed off the great runs of salmon, tend to be brown, and shockingly large when encountered up close.

Middle-Alaska bears are often black — their species name. They flow through the woods like panthers.

Far-north polar bears might as well be from another planet. Few people have seen these giants, unless you live in a village so far north you watch the ocean turn to ice each fall.

I am afraid of all bears. I am inferior in power, speed, and stealth, at sniffing and interpreting the scents of my surroundings, and in the ability to find daily meals in order to sustain a large mammalian body.

I have carried that fear for a lot of miles. But this time, the stainless-steel-and-plastic shotgun I gripped in my sweaty left hand 20 years ago was leaning in my closet at home. Its 7 pounds were too much. I walked and slept outside for an entire summer 20 years ago, and never once jacked a heavy slug into the barrel.

In 2017, I stowed a cannister of pepper spray, like a hand-size fire extinguisher, in the right thigh pocket of my convertible pants. It weighed less than a paperback. At night, I slipped that cylinder into a screen sleeve within the tent, always at my left shoulder. That way, I might recall where it was in a 4 a.m. stupor.

During the day, while walking, I rehearsed my OK Corral gunfight: Lift the can from the pocket with my right hand, flick off the plastic orange trigger guard with my right thumb, wrap my left hand around the base, point like a cop swiveling around a darkened doorway. Instructions on the label promised more than 30 feet of propulsion under ideal conditions (wind at your back).

When Lindsay's mom said all I needed for bear protection was Cora, she reinforced my logic in leaving the shotgun at home.

My walking partner possessed a magnificent sense of smell. Her ears perked to snaps of twigs. Cora's bark was piercing enough to cause pain, and has cost me some hearing over the years.

I was this confident in my little friend: At night, in the tent, I stuffed foam plugs in my ears. I turned off my radar.

The first week of the hike, before I tried earplugs, I lay in the bag listening to every noise out there in the twilight. My brain tried to make sense of nature's soundtrack: Was that splashing noise furred toes, crossing though the creek? Was the rustle of leaves an early-rising red squirrel, or something bigger? Was that rhythmic whack the wind slapping the tent fly, or footsteps? Each detected change in the outdoor symphony resulted in spike of adrenaline that further pushed back sleep.

Ear plugs worked. I relaxed, and slept through spruce cones that bounced off the tent.

Sometimes, though, I doubted Cora's effectiveness. Such as on cold nights, when she wormed her head into my sleeping bag. Or when she was so cooked from a day of busting through brush that her legs twitched to a doggy dream as I penciled in my journal. On those nights, I reasoned we would endure to smell the morning dew — we always had before.

* * *

Each evening, as my last chore I placed all my food — and Cora's saddlebags — in a trash compactor bag (unscented, please). I spun the bag and wound the twisted neck of plastic into a knot. I strung it between trees, or tossed parachute cord over the pipeline, then hoisted the bag up and tied it off.

The pipeline method was my favorite, for saving me the energy of shinnying up trees, and because it allowed me to lift food 10 feet off the ground in seconds. Up there, a bear might have trouble zeroing in on beef jerky molecules that escaped the bag. I don't know if Alyeska security people ever noticed the white bag laying atop the pipe, but I was careful to place it out of their sightline, and not to leave the bag there in the daytime hours that the security guys typically patrolled.

Tracks and bear scat on the pipeline pad were rare enough that I paused and took a photo of both prints and piles, with my shoe for scale. Forty years had passed since the intrusive, three-year construction of the pipeline in the late 1970s.

Those were the days when thousands of workers clomped over these hills, sometimes leaving their lunches on pickup tailgates. They fed bears when the opportunity arose, snapping photos with their new single-lens-reflex cameras, sharing the developed photos on their next shifts.

Those habituated bears were long dead, having nourished willows as their bones dissolved into the forest floor. Their offspring don't seem attracted to the gravel road running across Alaska. It holds little for them now, certainly not the ham sandwiches on white bread of 1978.

Today, security contractors wearing shirts with collars and handguns in molded holsters on their hips, creep along their trucks. And contract workers rise in cherry-pickers to replace the 40-year old ammonia with carbon dioxide in the ground heat-dissipating tubes. As I walked past at 3 miles per hour, I never observed those men and women throwing their half-eaten sandwiches in the bushes.

* * *

After a few days, my sleep improved as my body accepted the imperfect flatness of my tent sites, upon gravel or moss or leaves. The threat of bears in the night was a thought I brushed away.

Then one night I received a text on my satellite-tracking device.

"Did you hear about what happened?" Kristen wrote.

She went on to explain that a black bear had killed one of her co-workers. Within a few miles of where I'd soon be walking.

"It was a predatory attack. There was another one a day before, down by Anchorage."

Curious Ned of course wanted the details. The guy who was sleeping in the woods did not.

"Please don't tell me any more," I wrote back. "You can give me the details when I'm done hiking."

Kristen honored that request, but "predatory" sparked the dry kindling of imagination.

In the tent, my thoughts went to the black bear across the Melozi River. It was not deterred by a group of men on a small island, yelling and waving their bulbous packrafts. On that wilderness trip with four friends, I watched nervously as that black-as-night animal plunged into the river, swimming toward our island.

It turned back only when Jim Brader bounced a rock off the bear's head. Even then, after the bear turned, swam away, and then lingered on the far bank, it watched us without blinking its dark eyes.

Despite being exhausted from a long day, we climbed back into our dripping packrafts. We didn't camp until we found another island, miles downstream. As we floated, we swiveled, searching for a black head behind us, trailing a crocodile wake in the river.

There in my sleeping bag just off the pipeline, I pictured the Melozi bear watching us on the bank of the river. Its face did not show the slightest tick of fear.

* * *

My experiences were nothing compared to the terror faced by Kristen's co-workers. I read about the details after the hike.

The two biologists, working for an environmental research consulting business, were collecting soil samples not far from a major gold mine in Interior Alaska, close to the pipeline. Their work area was the boreal forest, the thick belt of spruce, aspen and birch trees that runs from Alaska all the way to Nova Scotia.

Many thousands of black bears had recently pushed their way out of hibernation dens and were padding over the landscape. One of the most adaptable creatures in Alaska, black bears eat everything from fireweed sprouts to composting salmon to newborn moose calves.

Like Kristen, the women — one 27, the other 38 — spent much of their time working outdoors in Alaska. They were comfortable there. The younger woman never imagined that day would be her last.

More than 30,000 grizzly bears live all over Alaska. They are not endangered; biologists find them from the tundra of the North Slope to the rainforests of Southeast.

Black bears are more than three times as plentiful. About as many black bears lope across Alaska (100,000) as there are people living in Fairbanks and its surrounding hills and valleys.

The younger of Kristen's coworkers was hiking ahead of the other on that June day, just before summer solstice. The birch were splotched with sunlight, in the time of no night. Swainson's thrushes sang their fluty songs. Mosquitoes whined in the cool air.

A black bear knocked the trailing woman down, from behind. She never saw it before the blow. Her backpack might have saved her, by absorbing much of the force. She heard a pop as the bear bit her water bottle.

She collapsed on the forest floor. Realizing this was a bear attack, she covered her face with folded arms.

For some reason, the bear then left that woman. It advanced on her 27-year-old partner.

The bear pounced on the younger woman. She screamed.

The black bear turned, advancing again on the older woman.

But now she had a weapon. During the initial attack, the woman's pepper spray had popped out of the rear pocket in her backpack. Despite her terror while hearing her coworker's wail, the older woman had the presence of mind to recover her pepper spray from the moss.

Now she saw the bear coming toward her. She aimed and blasted.

Orange dye painted the bear's face. A mist of oleoresin capsicum oil infiltrated its eyes, nose and fur. It turned away, blinking and licking at its paws. That bought the woman precious seconds in which to back off. She scrambled away, into the forest.

When she could no longer see the bear through the brush, she called for help on a satellite phone she carried in her pack.

She yelled to her young co-worker, saying her name over and over. No response.

A helicopter from the mine arrived. The pilot landed after scaring the bear away with the noise and wind-blast of the helicopter's rotor blades.

The pilot pulled the biologist who called him onto the helicopter; they left the younger biologist behind. The pilot later said the bear had appeared to be feeding when he arrived.

Hours later, a mine employee with a hunting rifle returned in the helicopter. The bear was again on the body of the younger biologist. The man shot and killed it. Alaska Wildlife Troopers identified the bear as a cinnamon color-phase adult male.

Experts determined the nightmare was a predatory attack.

"Bears may approach people out of curiosity, to test dominance, because they are food-conditioned or, rarely, because they are predatory," Alaska Department of Fish and Game biologist Doreen Parker McNeill wrote in a press release following the killing. "Interagency bear safety experts recommend people hiking, running or biking through bear country to carry a deterrent such as bear spray, or a firearm adequate for killing a bear. It is critical the deterrent be immediately within reach, where it can be quickly and easily accessed in the event of a sudden encounter . . . If the bear attacks, fight with anything you have, concentrating on the animal's face or muzzle."

Workers for the Alaska Section of Epidemiology analyzed that event, and all bear attacks in Alaska from 2000 to 2017. In a report they summarized, among other things, that the No. 1 victims of bear attacks were men of my demographic — white males between 50 and 60.

Here though, is what I found most telling: For all the anxiety bears provoke, the scientists concluded that hospitalizations and fatalities from bear attacks are "uncommon events."

During 2001-2017, Alaskans were 27 times more likely to visit the emergency room following a bicycle accident. When you add spark plugs to the equation, the numbers go way up: Four-wheeler and snowmachines sent people to the hospital 71 times more frequently than did bear attacks.

A grizzly bear once galloped after me on a riverbank as I tried (and succeeded) to outrun it in a motorized canoe.

I was once alone inside an abandoned building with a polar bear. When I noticed its fresh-plate size tracks in the snow that coated the floor, I was unable to move for a few seconds, as scared as I've ever been as an adult. Wincing at my boots squeaking the cold snow, I retreated out of the building thinking each second would be my last. I never saw that polar bear, but no tracks were outside that building — the bear was inside with me but decided not to show itself.

After those encounters, I have lived many more seconds. Those experiences were outliers in my hours spent outside four walls in Alaska.

Those, and other bear memories, have faded amid hundreds of hours in quiet, peaceful woods. Even though I wear earplugs in the tent, deep-down I know I am safer on the muskeg than I am while driving.

Kristen, who has feathers of white skin on her shoulder from the swipe of grizzly claws, as well as a crescent scar on her wrist from the same bear's fang, has valid reasons to not sleep in a tent. But she still does. I will leave her story for later.

As for me, I don't think a bear will write the final page of my journal. I am scared sometimes, but still I value nights sleeping outside more than those in my bed. I don't let fear prevent me from driving across Fairbanks to play softball. Both the driving and the slides into rubber bases subject me to more risk of injury than do muscular, hairy beasts born in mountainside wombs.

Just like the guy who buys his gun at the box store and one day lets loose, rogue bears get all the press. The headlines ignite our primal fears of dying by tooth and nail, so deeply wired within us. Living inside for many generations has not erased this programming.

But our guns, and tubes of pepper spray, cause much more misery in accidental firings than bears ever will.

For me, after again spending an entire summer living outside in bear country, it all came down to sample size. For 120 days in 1997 and so far a few weeks of 2017 — as well as the 7,000 days in between — I had avoided a bad bear encounter.

I try to do what I think is best: I am fastidious with my food, sing poorly in thick brush and shadowy terrain, and travel most often with a dog on voice command.

Despite all that, I have concluded that my clean record is attributable to the extreme low probability of a bear attack. And dumb luck.

5

The Big Lonely

Most Alaska non-fiction books include a Bear Chapter. You've read it before. Sorry.

Bears, though, are part of the Alaska Difference. Balls of muscle coated with rippling fur tickle the neurons of your brain during every Alaska trip, even if you hike the same path 20 years later.

In the two decades between my walks, the world population of us two-legged creatures had increased by more than a billion.

The explosion was a function of human biology — sex feels good, babies appeal to us, and nurturing a fledgling to adulthood gives us purpose. The Bible urges us to be fruitful, multiply, fill the earth and subdue it.

There are other factors that helped Homo Sapiens. We planted things that grew, so we wouldn't have to chase all our food down. We were clever enough to invent eyeglasses, and medicines that make our flaws less fatal.

Along the way, we were no longer satisfied with a den dug into a south-facing hillside. Instead, we drove by a woodlot with a nice view of snow-capped mountains. We noticed a "For Sale" sign hanging above the berry bushes. We imagined sitting on a deck after dinner with a bottle of beer, dreaming on those rugged peaks. One noisy summer later, the sawdust settled to black earth, we smelled the kiln-dried Oregon lumber of our own Alaska Dream.

To avoid the new structure, a black bear veered from the path her mother showed her. Now a mother herself, she led her cubs along a narrow buffer around the driveway, searching for a new blueberry patch, far from the newest intrusion.

But the pause button is being pressed here in Alaska. The human population is no longer increasing.

Cora and I saw it in the U-Haul vans, transmissions groaning up Thompson Pass. People driving those boxes on wheels, squinting at the road ahead, were gone for good. Aiming for the Tok Cutoff Road. Heading for America.

Following our encounter with the residents of 46-Mile, Cora and I walked together into open alpine country, mountains always at our shoulders. Eventually, we descended a long hill and back into the boreal forest that wraps middle Alaska in a dense, sprucy belt. There, I hoped to find my food stash still tied in a tree.

Grabbing sticky, fragrant branches thick as a broomstick, I climbed high into a spruce and reached for my plastic compactor bag. I had secured it there nine days earlier, on the drive to Valdez with Kristen and Anna.

I tugged to release a slipknot of parachute cord and lowered the heavy sack 15 feet to the ground, thankful that ravens had not been attracted to the white plastic perched in a fan of green branches.

After spilling the week of supplies on the forest floor, I re-stocked Cora's saddlebags with doggy nuggets. Into my pack I slipped foil dinner packets, trail mix, my own recipe of oatmeal with dried cherries, and — most important — happiness-inducing Starbuck's Via coffee singles.

The heavy pack pushed down like two meaty hands on my shoulders, triggering a feeling of security and well-being. Without my coffee packets from California and dried cherries from Michigan, I would be lost in the hungry country in which bears somehow find their fat.

Resupplied, I hiked on. I soon felt a few sparks of apprehension, due to the bushiness of my alone hours with Cora. I anticipated human encounters just ahead. Less than one mile away was Pump Station 12.

About 700 miles from Pump Station 1 in Prudhoe Bay, Pump 12 is the last of a series of permanent camps built by Alyeska Pipeline Service Company workers to propel arctic oil across Alaska's bumpy face to tankers in ice-free Prince William Sound.

In addition to their hydraulic transportation duty, pump stations once housed hundreds of workers. The compounds accepted hard-hatted workers back after cold and wet days in the field, fed them steaks from Nebraska, and felt the slams of dormitory doors as smiling men and women departed on leave, hefty paychecks in their wallets.

The heart of the futuristic complex at Pump Station 12 is a building the size of a bus garage. Just before entering the station, the pipeline dives beneath the ground to penetrate the structure. Three turbine engines there propelled the oil southward, giving the crude a final shove over Thompson Pass so it could reach the finish line at the Marine Terminal in Valdez.

* * *

As Cora and I approached the pump station, an old friend appeared at my right. The shiny, four-foot thick python of trans-Alaska pipeline had — for the first time on our south-to-north journey — popped above the ground to avoid the permafrost engineers had long ago detected in this valley.

I was happy to see the pipe. It would allow me to hang my food easily in the coming nights. Scattered, handwritten notes, written decades ago in welder's pencil on support beams, would give me something to read, a few cryptic messages to think about as I walked.

Something else caught my attention as we neared Pump Station 12. Twenty years before, the roar of those jet engines echoed off snow-dusted pyramids of mountain. This time, I heard the friendly calls of chickadees and the wind, whispering through spruce. No industrial hum washed them out.

I clipped Cora to her leash as we approached the gates of the pump station, because I didn't want her to run up to someone who was packing a sidearm and not expecting a flash of black.

I needed a wi-fi signal to send my weekly column back to the Geophysical Institute, where a coworker would edit it and send it to Alaska newspapers. I approached the guard shack at the entrance of the complex, but saw no silhouette of a security officer.

The shack was empty.

I noticed a phone box on the chain-link fence that

surrounded Pump 12. I walked over, picked up the receiver, and heard a ring.

"This is Jeff, Alyeska Security," a voice said.

"Uh, hi. Do you have wi-fi here at the pump station?"

Jeff laughed.

"No, there is none," he said from his Anchorage office. "Pump 12 closed a few years ago. You're in the Big Lonely, man."

* * *

Earth's human population increased from 5.9 billion in 1997 to 7.5 billion in 2017. For a long time, Alaska's numbers had risen along a similar curve, but in 2017 they were trending the other way. Just a year before, the state's population had decreased for the first time since 1995. Why?

State demographers say fewer people are moving to Alaska, and Alaskans are having fewer children. Department of Labor economist Neal Fried, who has been at the job for more than 30 years, also said that low unemployment rates in the Lower 48 states have kept people from packing up and moving to Alaska.

"If you can find good economic opportunities closer to

home, you're going to stay closer to home," he told Anchorage Daily News reporter James Brooks.

Many long-time Alaskans came up here to work for a summer but stayed for a lifetime. Fewer people are now taking that first leap.

Alaskans are packing U-Hauls and driving the only road out of the state. Many have said goodbye to Alaska for the same reason Pump Station 12 was quiet — much less oil is being extracted from the frozen ground, and lower oil prices have kept the rest of it beneath the Labrador tea.

* * *

In the book "A Land Gone Lonesome," by Fairbanks's Dan O'Neill, the author takes readers on a canoe trip down the Yukon River, from Dawson City, Yukon, just east of the Alaska border, to Circle, Alaska, a community about 150 miles northeast of Fairbanks.

Both are Gold Rush towns that once boomed with dusty miners walking plank sidewalks through plumes of perfume, with piano notes floating from gambling halls.

The book's title suggests that the country surrounding O'Neill on the flat, brown river is less peopled today than it was in those days. Gone are the log roadhouses, built to feed and shelter travelers every 15 or 20 meandering miles. Also absent are the river people of the 1970s, mostly young white people from the Lower 48 who had learned to cut salmon and hang them on spruce poles in a warm, dry breeze.

The river people often left that country because their Bush-born children — more skilled at wilderness craft than 99 percent of other Americans — were entering their teen years. Despite their unique upbringing, they were still teenagers craving others their age. Their parents also sensed an upcoming loss of freedom. The land upon which they had built their cabins belonged to someone else, either the federal government, or the state of Alaska, or Alaska Native Corporations. Officials of those entities were starting to inventory their assets. As they became aware of the river people, officials with those regulation-driven entities began to expel the low-impact intruders.

Their cabins, now sinking into the muskeg, are out there waiting to be discovered by travelers who explore the clear fingers of river merging with the Yukon.

* * *

Twenty years earlier, pipeline workers advised me and Jane to avoid the pipe's route around a low contour of 4,000-foot Willow Mountain. Willow Mountain is the Chugach Range's last shoulder leaning into the Copper River Valley.

"People will pull a gun on you," the workers had said of locals with land just off the pipeline right-of-way.

I heeded that advice. Jane and I walked a 16-mile day on the shoulder of the highway to avoid Willow Mountain. It was a long day of yanking Jane back off the road to avoid whizzing cars and trucks.

In 2017, I wanted to see the section I missed. Instead of taking the access road to the Richardson Highway, Cora and I stayed on the pipe's path, around the base of a mountain that still felt ominous to me.

I held my breath and leashed Cora. I soon came upon a driveway leading off the pipeline pad. I imagined the hillbillies that the pipeline workers had warned me about 20 years earlier. I thought I could sense being centered in a rifle scope.

But a closer look showed fireweed sprouting in the wheel ruts of the driveway. No scent of spruce smoke wafted to my nostrils on that chilly morning.

No one was home. Nor had they been, for years. It was the same for the half-dozen other places I walked past beneath Willow Mountain. The little community had been abandoned. Old trucks sat decomposing into the moss, sheet metal twisted to ghostly shapes as it peeled from roofs, driveways sunk into the ground, with no one there to dump in fresh gravel and rake it level.

Willow Mountain had gone lonesome. The vapor of Alaska dreams that once hung there had drifted away, to Seattle or Santa Barbara or Little Rock. Or maybe Heaven.

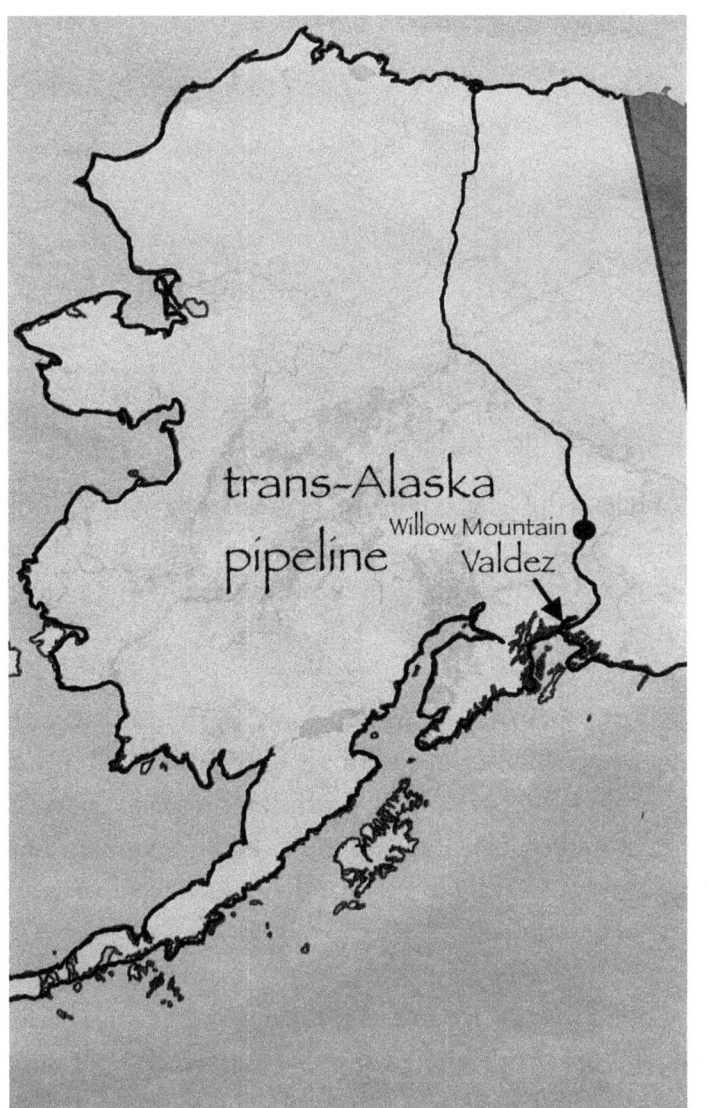

6

Spark plugs

While I drove down from Fairbanks to start the hike, Kristen, Anna and Cora rode with me. When the girls and Cora dozed to the hum of the road, a Need-to-Get-Done list tumbled through my head.

My trip preparation wasn't done. In the car, we carried along a few white compactor bags filled with food, fuel cannisters, and Ziplocs bulging with dog nuggets. These food drops would enable Cora and me to keep moving through a large swath of southern mainland Alaska devoid of grocery stores or gas stations.

Country without commerce is a major difference between this 800-mile path and established long trails that stretch across the Lower 48. On the Appalachian Trail and Pacific Crest, towns and villages appear at handy intervals. Hikers there at tick-and-snake latitude can fill their empty packs, stuff their bags of trash in a dumpster, and catch a shower every three-to-five days.

Not so, at least so far, on the path of the trans-Alaska pipeline. While the hamlets of Glennallen and Delta Junction give drivers reasonable distances to fill up their tanks between Valdez and Fairbanks, the hundred-mile-plus spreads are too far for a walker, at least one with my pace.

I can carry a week's worth of food and fuel. More is too heavy, and too bulky for my featherweight, single-compartment ULA Catalyst. A three-or-four-day load is just right for maximum happiness. That way, the minimalist pack is comfortable enough. After a few hours of walking I do not even notice it back there, except on restock day.

To enable restock day, I needed to hide those plastic bags amid the acreage of the Big Lonely.

* * *

In 2017, I had fewer options to resupply than in 1997. Tiekel River Lodge and a gift shop at Pippin Lake were closed.

Thus, the bag I had tied in a spruce tree, and the one Kristen and Anna had secured to the girders of a highway bridge for me. I had also tucked a sack in the Alaska Range cabin of a friend.

I thought of another possible stash site near Copper Center, a riverbank town that had shrunk considerably since the Gold Rush of 1898. Back then, a few hundred men had stumbled in over the mountaintop from Valdez, after hiking a few glaciers and crashing wooden boats down a raging river.

A little outside Copper Center, Mike and Lanette Phillips live on a lovely aspen-shaded bench above the aquamarine Klutina River. I had encountered them 20 years earlier, on my first hike, and had featured them and their two kids in *Walking my Dog, Jane*. I had not seen them since, except for a book signing in Glennallen, three years after the first hike.

But here I had a need — to drop a bag I could walk back to in a few weeks. The Phillips' homestead would be a logical place, about four days north of the bag tied in a spruce tree. They probably had a shed in which my food would be safe from bears, squirrels, dogs, and ravens.

Someone better at planning would have pinged Lanette on Facebook a few days before he came down. But contacting the Phillips was one of the details I did not get to; it was time for a cold call.

* * *

A day before the hike started, I turned our Honda onto the

gravel road that led over the buried pipeline. The car groaned up the hill the led to the Phillips' house.

No one was home. The house was as I remembered it — lovely, neat, surrounded by aspens, their crowns swelled with buds that hung like sticky pendants. I did notice a difference. A wheelchair sat on a ramp, at the top step of the porch.

Because no one was at the house, I reversed course in the Honda, driving back down the hill. We motored a loop around Copper Center, an oval that did not take long to complete. We noticed that both the lodge and the store were closed.

"Let's give the Phillips another shot," I said to Kristen.

We turned and headed back up the hill.

This time we were in luck. A pickup sat in the driveway, engine clicking as it cooled. Around the back of the house, I noticed a thin man wearing sweatpants, sitting on the seat of a four-wheeler. He watched me walk toward him.

Mike Phillips said hello. I recognized his penetrating blue eyes and deep, resonant voice.

I waited for him to get off his four-wheeler. When it was evident he was not going to dismount, I walked over and shook his hand. His face fired with recognition.

"Well, look who has returned," he said, smiling as he shook his head slowly from side to side.

* * *

A few weeks later, after 14 days of walking, Cora and I again neared Copper Center. After many nights of sleeping on the ground, we had softened up with a few indoor nights.

The first of those evenings in soft, clean beds came with the aid of my satellite texter.

Squinting at the little screen in my palm, I asked my friend in San Francisco if he could find an email for a friend near Kenny Lake, south of Copper Center. After a few sentences had traveled 22,000 miles to a satellite and bounced back and forth a distance equal to a few trips around the Earth, I once again walked into the 5-acre homesite of Doug Vollman.

Like the Phillips, Doug was also a character in *Walking my Dog, Jane*. I met him during my 1997 hike as he tended bar at Tsaina Lodge, just north of Thompson Pass. He invited me to his home 60 miles north of the lodge, and to play golf on the nine-hole course he had laid out on his muskeg acreage.

In meeting that youthful spirit back then, I felt a faith-in-humanity blast — a stranger who fed, housed and entertained me simply because I was a fellow human who had brushed against his life. Interacting with the owner of the Ford Torino hazard on hole No. 5 convinced the writer in me there was a book in my summer trek — not just about a guy walking with his dog, but a snapshot of the people who occupied the imperfect land along the way.

Upon my return to Doug's homesite, this time with Cora, Doug and I did not play golf. Other projects, like Doug's growing vegetables for a local farmer's market, had taken his hours. Wild rose bushes had crept in on the golf course.

Things had changed in 20 years. In Doug's home, Cora enjoyed belly-pets from Doug's daughter Taylor, back at the family cabin after a semester at the University of Alaska Fairbanks. A few weeks earlier, Taylor and I had both been stretching on a carpet within the Great Hall, enrolled in the same yoga class. Before class, I would ask her how her parents were.

When I had visited 20 years earlier, Taylor, now 19, was just about to be sparked to life. Here, she flashed her mom Marnie's happy smile, a miracle duplicated. Marnie was not at the house. She was in Jackson, Mississippi, for a two-month training course for her job with the federal government.

Doug and Taylor

Doug, in 1997 a spirited 40 and now an energetic 60, whisked me into his greenhouse, attached to his main house. He whirled his arms over legions of vegetable starts, their blades waving back at him. I inhaled deep of their exhaled oxygen.

Back inside, Doug turned on the TV for a San Francisco

Giants game. He fried me two cheeseburgers. Seeing them disappear, he asked if I wanted a third. Yes.

I stayed the night, with Cora on the bed, in a spare room. As I looked around in the twilight before slipping into my sleeping bag, I saw in the bookcase a yellowed, signed copy of *Walking my Dog, Jane*.

It seemed a lifetime ago since I was there last. It was. Taylor Vollman was living proof.

* * *

Recharged by human contact, I returned to the pipeline pad the next morning. Hours later, after a crisp May day of sunshine and the big voices of thumb-size ruby crowned kinglets fresh from Venezuela, I hiked my way back up the hill to Mike and Lanette Phillips's place, and to the bag of food I had left in their shed.

I walked their road, which was lined on both sides by purple flowers that seemed to wave at Cora and me. Their lavender was a neon contrast on the early-spring ground of brown and tired green. I remembered those early blooming crocuses had greeted me in the same place 20 years earlier.

Mike was there. He was outside, again seated on his four-wheeler.

"Ned made it back!" he called out.

Cora, outfitted in her backpack, ran over to his four-wheeler and put her front paws on his left leg. Mike reached down and petted her. Lanette walked out on the porch.

"We've been reading your stories," she said.

In 1997, I had described Lanette as "a youthful brunette with white hairs framing her face as if she were outside walking at 30 below." Two decades of senescence had transformed that hair into salt and pepper, mostly salt. Her impish smile — the one that puffed Mike with pride when other pipeline workers noticed her on his arm in 1978 — remained.

I dropped my pack on their lawn. Mike motored over on his four-wheeler. It was obvious then that he could not walk. He noticed that I noticed.

"I'm a dead man walking," he explained. "Heart attack, ulcer, septic infection. I go from bed to wheelchair to four-wheeler to truck."

I did not know what to say. I thought back to when my Mom was diagnosed with Alzheimer's. And when the family doctor gave me the prognosis on my Dad's cirrhosis of the liver.

"He's gone beyond the limits of recoverability," the small-town New York doctor told me on the phone.

That didn't allow space for much hope. At the Phillips', I again felt that silent despair.

But in the heaviness was also joy — of me reuniting with two kind people at their uncommonly beautiful place, the snow-covered Wrangell Mountains filling the eastern viewshed.

Mount Wrangell, a shield volcano, resembles a world leader lying in state under a white sheet, right down to the point of his nose. There had been no resurrection; none of the volcanoes of the Wrangells had erupted since I last sat at the Phillips' outdoor fireplace. No mountain explosions, no asteroids crashing into the planet — more reasons to be thankful for pleasant reunions.

* * *

As I had 20 years before, I slept two nights with the Phillips. This time, rather than pitching my tent on the lawn, I rested in the bedroom in the basement they had offered me last time.

On an end table sat a dusty portrait of Lanette as a young girl — smiling as if pulling a prank — with her sister and mom. Also, photos of the Phillips' children. Daniel and Colleen, who had fledged and left Copper Center.

Before I unstuffed my bag on top of the mattress, I listened. I heard nothing in the house, except for a few creaks on the floor above my head. The Phillips' empty nest was a quiet place, with the whisper of prayers. I had maybe recited the same ones since the last time I slept there, asking a God I wasn't sure of to reduce my parents' suffering. I there said another, asking that my hosts might find a bit of the serenity they had again provided me.

Lanette, Mike and Ned

While designing a pipe that would span America's largest state, including hundreds of rivers and streams, engineers chose different ways to cross the water. Excavator operators dug a trench under the frozen Salcha River so the pipe could be buried beneath it. Welded sections of pipe proved rigid enough to span Fish Creek in a 40-foot section. Each waterway featured a custom solution.

The pipeline above the Tazlina arcs over the wide glacial river with no bridge beneath. Instead, steel cables attached to towers on both ends of the river suspend the pipe over the river.

The only way I could cross the Tazlina there was by boat, unless maybe you walked on top of the pipe. I once considered that, but I doubted my dog partner had toenails sharp enough to grip the aluminum skin. Also, I imagined there were a few cameras pointing at the pipeline there. I did not have permission to tightrope-walk the pipe.

* * *

Mike four-wheelered down to his garage. I walked down. Cora followed me. There in the cool shadows, he had a boat, its two pontoons resembling giant bananas.

I helped drag all the parts from the garage. A tubular metal frame in the shape of a square supported a seat in the middle.

"Let's inflate this and see if it might work for you," Mike said.

I was game, though I did not ask how we might get

this 100-pound craft to the bank of the Tazlina. Mike read my mind.

"I think I can get this on a trailer, pull it to the crossing with my four-wheeler," he said.

As his air compressor hummed to fill the air bladders of the boat, he shared a sudden thought.

"Wait. This is a bit much, isn't it?"

"What do you mean?" I yelled over the white noise.

"Spark plugs. Too many spark plugs needed to get this thing to the river. It's not your style."

Before I could answer, the boat made the decision for us. The right bladder hissed with a leak.

"Let me make a few phone calls, try my good-old-boy connections," Mike said. "We'll see if someone has a smaller boat, one you can carry."

Twenty years ago, when I met Mike, he was riding a three-wheeler (since dinosaured by the more-stable four-wheeler). He then had said he understood when I declined his offer for a ride up to his house, or for him to carry my pack on his rig.

"I don't want to mess up your system," he had said.

He was perceptive enough to recognize a slight difference between us. After a half century of life, I have concluded that I'm one of those people who are not comfortable with engines.

I have a complicated relationship with spark plugs. Chainsaws are noisy and stinky and can send me to the Emergency Room in one careless drop of the blade, but I love dropping trees and bucking them up. Cars and trucks and 737s show how clever we are — converting a volatile liquid to controlled explosions, a few thousand of which propelled me to the start of the hike.

But, as we have learned, emissions blown from these amazing machines have hastened the warming of the air blowing around our home planet. That has, among other things, helped thaw the ancient frozen ground surrounding my house.

The structure has slowly sunk on its post-and-pad hybrid foundation. Over time, it squatted with all its force on our rigid plastic sewer pipe, where that tube dove straight into the ground. The black pipe cracked one November day. For a few days, before we knew what had happened, we wondered what might be the source of the liquid squeezing up from between seams in the birch flooring.

Another observation from 30-plus years in Alaska: Though

I have walked their rolled-in paths of least resistance many times, four wheelers are a most unfortunate invention regarding the backcountry. They have allowed us, for $5,000, to replace our boots with four unstoppable knobby wheels.

Alaska is not a cold, hard land. Much of this rugged country is delicate as a pile of sponges. In a few minutes, four wheelers can spin out on wet greenery, tearing out dwarf birch and miniature Labrador tea by the roots. Erasing decades and centuries of patient growth.

The black muck left behind absorbs the warmth of sunlight. With the insulation of the plants stripped away, warm air touches soil that has been frozen since the days of the mammoth. Left behind is a boggy mess, no good for four-wheelers or feet. We skirt around, braid the trail wider, spinning new webs.

The person pressing the gas lever with his or her thumb probably did not intend to destroy the tundra. He or she just wanted to get somewhere. It was fun to get through to the other side of the mud bog. Or maybe it was just necessity to get from one place to another. Either way, the damage is done.

Four-wheelers have their place. Mike uses his where a wheelchair doesn't work. He drives it down his gravel path to the Klutina River. There, he sits and watches the emerald froth and listens to the roar. Miles away, on pebbled village streets all over Alaska, four wheelers move water, firewood, and the greasy bulk of bearded seals. They are fine on rocky, solid trails and gravel roads. But too often we take it too far, because four-wheelers can make it through.

The next morning, John Rigo, a member of Mike's good-old-boy network, rolled into the driveway in his rusted blue pickup. He parked on a gravel plateau by Mike's garage.

The former smokejumper stepped out of the truck and shook my hand, so firm that I winced. He looked like me in about 10 years, a bit grayer, but still getting it done. John Rigo then pulled from his cab a blue nylon bag not much larger than a loaf of bread.

That small bag covered a packraft, made by the Alpacka Raft company, born in Alaska but now operating out of Mancos, Colorado. Sherri Tingey, a fabric designer, started the company when her adventuresome son returned home from an Alaska trip with a shredded boat. Later on my trip with Cora, Sherri would loan me a similar raft. It weighed less than 5 pounds.

I reached into my wallet to pay John for the rental of what was the perfect tool for crossing the Tazlina: Light, buoyant, no fuel required.

"No, just give it back to me when you reach Glennallen," he said.

The next evening — after a day of hiking with paddle blades projecting like antennae from my pack — I inflated by mouth the raft on the rocky shore of the Tazlina River.

Then, in mid-May, the nights were still cold enough that the river was not fizzing and boiling, as it would be in midsummer, when the glacier upriver would lose much of the snow-weight it had gained in winter.

I slipped the 5-foot-long craft into the water, my pack listing in the bow. I sat in the back and called Cora over. A fearless veteran of many canoe rides and a few unplanned swims, Cora jumped in and nestled right between my legs. Perfect.

Pointing the bow upriver, I ferried across the river at a 45-degree angle. It allowed the current to move us across without much effort from me, other than dipping the kayak paddle to maintain the slant. I didn't worry about the current pushing us downstream a bit. After a few weeks of walking, the water's velvet hand was magic.

As I crunched into fist-size gravel of the far bank, Cora leapt out.

I adjusted for her push, then again paddled the bow onto gravel. My boot found a rock and I pressed myself on the tube to push up and out of the packraft. A snazzy dismount. Though I usually emerge from those boats with a dripping wet butt, this was a pleasing dry exit.

The night was sunny and warm, smelling of riverside sage. I found a campsite amid nodding Einstein heads of dryas plants, on a bench overlooking the Tazlina River.

The river's flow was a song that would not pause until the river froze in November, months that were hard to imagine as high summer approached. Other than the gurgle and hiss of the water, there was no sound of distant motors, though the liquid that would fire spark plugs from Topeka to Tumwater was sloshing through the pipe just 100 yards away.

I pitched the tent without the rainfly, so I could watch the river flow by as I was cozy in my bag. Cora pressed warm against the back of my legs. Soon, she starting snoring. That soothing rhythm, and the whisper of the nearby river, closed my eyes.

7

A lot like me

"I'm never coming out here again to hike the pipeline," Anna said. "You made a bad decision."

My 10-year-old said that to me, as we walked together between peaks of the Alaska Range.

At the time, it was hard to debate with her. Forty mile-per-hour winds shoved us, drilling raindrops into our cheeks. We could not keep our eyes open for the bullets of pain.

My girl is good at arguing. I tell her she would be a good lawyer, though I hope she does not pursue that line of work.

Two girls, Anna and her friend Salak Crowe, were hiking the path of the pipeline with me and their moms. For nine days, they joined me through the Alaska Range, from Meiers Lake to Black Rapids.

Those 60 miles of trail featured the worst weather Cora and I had experienced so far in our one month of hiking. Sub-freezing temperatures each night. Rain, wind and graupel pellets during the day. One morning — on Memorial Day in the Alaska Range — we woke to three fresh inches of snow.

We adults were freaking out a bit. How are we going to keep the girls warm? How could we keep their spirits up and their bodies moving forward, to the car waiting at Black Rapids Lodge?

The first three days, we broke the Golden Rule (no food in the tent!), handing the girls bowls of oatmeal through the tent flap. We fretted about hot chocolate spills on sleeping bags and the tent absorbing bear-attracting odors.

But when it came time to hike, the girls popped out of their tent with boots on and backpacks full. One of the adults, set to go and carrying bear spray, would get them striding down the trail to warm up.

And there, after an hour, was one of my favorite images: the girls, leggy as newborn moose calves, walking hip to hip, talking, singing, never running out of things to say. Anna walks with a bounce in her step that reminds me of my younger brother. Salak has a light, pigeon-toed stride that looks like her father's. Their smiles and happy chatter warmed my wind-chilled heart.

Anna

There were other times, though, when Anna whined at the wind and her cold feet. She vocalized exactly what I was feeling, but had learned not to speak as part of my Alaska hardening. Instead of responding by yelling, I had to walk away sometimes, and let Kristen take over.

Our girl does not "suffer in silence," something my mom often requested of her five children. Anna is a lot like me, which maybe leads to greater understanding, but also results in a lower tolerance of each other's impatience.

* * *

During nine days of Aleutian weather, the girls impressed us adults. Within the 20-year-old tent they shared, Anna and Salak stayed up for hours, chatting and giggling. It was mountain music.

Maybe there is something to growing up Alaskan. River trips with real hazards of bears and icy water with rocks. Outdoor recesses at 20-below. Mosquitoes, mosquitoes, mosquitoes.

I didn't have those elements growing up in a New York mill town. But my parents did mobilize the five kids for camping trips to the Bertrands' upstate property, and somehow to Maine in a Volkswagen bus. There the outdoor seeds were planted.

This raw, open-air world has also been thrust upon Anna and Salak. Regardless of where they end up (Anna says she likes Brooklyn, where her aunt lives), the girls will be shaped by this oversize peninsula.

Salak, known as "Sala," placed rocks on her mittens so they wouldn't blow away as we filtered water from a creek. Anna flicked the lighter to fire the camp stove. They teamed to pitch and take down the tent. Their hiking pace was as fast as us adults, and they weathered the elements with less alarm than we did. We must have been good at hiding our panic as our eyes darted to the girls as they leaned into the sleet and tested their rain jackets.

Hiking alongside Anna, I told her I wanted to spend time on the trail with just her and Cora later that summer. I did not want to force her along, but I thought time alone with her would be fun for both of us. We have a rapport together that is not the same when her mother is along. I wanted some of that vibe in a summer during which it totally made sense because she was not going to school.

"No," Anna said. "I'm not coming back out with you. This is your trip, not mine."

I did not let her see me wince when that dart hit my heart. But neither did I argue with her logic. I clinged to a secret hope. Just like me, "no" is often her immediate response to a proposed plan. She needs a while for a notion to slow-cook Sometimes, she changes her mind. I do. Sometimes.

SUMMER OF GRAVEL AND STEEL

Often, she does not ever agree with the plan, especially those involving human-powered trips to cabins on packed snow trails. Kristen and I have struggled with Anna having proclaimed herself an "indoor girl."

Once, after hearing that her mom signed her up for a race, Anna said "You're wrecking my life with skiing!"

We hoped that she will learn to love living outside as we do. So far, we are not seeing so much of that desired evolution. But we are coming to grips with it. She's right: it's her life, not ours.

Some kids inherit the same blueprint for joy as their parents, some don't. Each wriggle of sperm into egg results in a random spirit, with a likelihood — but not a guarantee — of similarities to the parent organisms. It's a crapshoot, necessary for a species to evolve, as some differences become advantages.

Anna's indoor nature seems not so bizarre during our deep, dark, half-year-long winter. Below about zero Fahrenheit, the air hurts your nose and fingers and anything else not wrapped. That frigid condition exists from late November until February in Fairbanks. It takes a bit of extra punch to bundle up and enjoy the boreal forest in midwinter. It's a bit of a push for a species that emerged near the Equator. Those of us who enjoy winter are outliers.

So, maybe Anna has something there. Or maybe she does

not have the restlessness that pushes me to pull on my down pants and jacket for a dog walk at solar noon in December. She seems fine with watching me go out, saying "I'm OK," when I ask her to come along. And OK is all right. I guess.

* * *

A few days later, with the wintry conditions persisting, I said goodbye to Kristen, Jennifer, and the two girls, in the parking lot of the old Black Rapids Lodge. The girls were snug in back seat of the car as a breeze off the glacier across the valley rocked the chassis crazily.

Just before I turned to continue down the trail, the back door swung open, the wind almost ripping it from the hinges.

Out ran Anna. She jumped in my arms for a hug.

"So, will you come out and see me again?" I asked.

"I'm thinking about it," she said, squeezing me before climbing down. "Love you."

8

Birds

Water is life. The proof was in the willows and the mosquitoes and the dragonflies and the gnats. As I sat on a streamside pile of gravel shoved into place 40 years ago by men driving D-9 Cats, I was in a biological hotspot.

While stirring my blue-glowing Steripen in the water bottle for 90 seconds, I watched small bodies flit at eye-level in webs of streamside branches. Flashes of tropical color, often lemon yellow, revealed the almost-weightless flesh and feathers of warblers, thrushes, finches and so many other songbirds that had just returned to Alaska.

Several times every day, I paused and made water safe to drink.

A yellow-rumped warbler

I stopped walking when I reached a stream that cut through the pipeline pad or flowed just beside it. I dropped my pack, feeling the phantom rise of my shoulders toward the clouds, and then clicked twice with my tongue. Cora appeared, knowing I would then release the two straps of her pack so she could step out of it. She, with her infinite energy, would then wedge into the nearby shrubs.

Sitting on a flattish rock if I could find one, I reached into my small gray bag of miscellaneous stuff and pulled out the Steripen. That innovation, which had emerged since 1997, enabled me to shed weight. The Steripen purifies water by blasting it with ultraviolet light for 90 seconds. The size of a thick candle, it was one-fifth of the heaviness of the plastic-jacketed ceramic cylinder through which I pumped my water 20 years earlier.

In addition to keeping me hydrated, the water stops stationed me in what my wife and other biologists call the riparian zone. Derived from the Latin *ripa* — "river bank," the landscape along creeks and rivers provides insects and tangles of good nesting material for birds.

Nestled along a watercourse, I was never alone. With Cora off trolling the brush, often it was just me and a yellow-rumped warbler, perched a few feet above my head, its dinosaur feet wrapped around a willow stem.

I felt a pure, ephemeral blast of happiness. Those flicking, palm-sized creatures, which only remained in my viewshed for a few seconds, were a miracle revisited. Like the return of the sun to Interior Alaska after winter solstice, the birds fulfilled a promise as they flapped and glided back to us from winter perches on tropical tree branches.

I was in the midst of a flesh-and-blood-and-feather flood, flowing northward. Sometimes I could sense it. Most times, I didn't think of the billions of beating hearts returning to Alaska. But they were coming: blobs on weather radar that sometimes caused me to look upward from a nighttime trip outside the tent. I'd squint at the faint honks so far overhead.

All around me, birds were touching down on branches and brushing their waterproof chest feathers onto the cold liquid of lakes and rivers. As birds have every spring for millennia, they returned to a place awakening from the dark, quiet sleep of winter. A few hardy residents, among them the ravens that sometimes seem to follow us by leapfrogging pipeline supports, make Alaska their home all year.

Migrants transform Alaska from a one-stoplight town to New Delhi each spring.

Some, like crow-size whimbrels — their curled beak like the stem of a Sherlock Holmes pipe — were landing on reindeer moss, the first soft earth their toes had felt since lifting off the desert sands of central Mexico. Those birds had bypassed both the Lower 48 and Canada — on the wing for three days and nights without eating or drinking — to reach to a tundra bench above the Kanuti River.

Why go to all that trouble? Why do so many birds travel global distances to reach Alaska?

To quote my friend and bird biologist Susan Sharbaugh (who texted me Yankee scores every night), the birds were coming back because the kitchen was open after so many months of darkness. Bugs, wonderful bugs!

Look at a satellite image of Alaska and you will see some of the planet's greatest swamps. Wetlands of this extent are absent in Europe, where scientists have documented an alarming recent decrease in insect populations. Biologists have not recorded that same dearth in Alaska, nor North America as a whole. That's good news for the birds, and people.

In one of those Iowa-size bogs, the spruce-and-tussock lowlands of the Copper River valley, I had a hard time sleeping with the sudden stirring on the landscape. Camped between two ponds underlain by frozen ground, I woke several times to the croaks of wood frogs, the protest honks of swans, and the chuckle of ducks. Twice that night, moose clicked right by the door mesh, one so close I could have hit his nub antlers with a tossed water bottle. As this world thawed, the season of biological productivity — and reproductivity — was underway. I could almost smell the love.

All those birds, some larger than a Thanksgiving turkey and a few smaller than your thumb, return to Alaska because of the Yukon Flats, the treeless North Slope, the lumpy Seward Peninsula and the great, sodden plain of the Kuskokwim River.

With few cities and 240 low-impact dots of Native villages spaced along the river systems, Alaska has vast, mostly intact natural systems.

Wet muskeg that can swallow an F-150 is perfect for hiding from a raptor that might eat you or your eggs. Streamside alders with upturned stems feature solid wood forks that hold a nest of woven grass, invisible to all but the parent birds. Cliffside ledges above rivers are inaccessible to foxes and coyotes, but not to peregrine falcons that return to the same white-stained rock platforms each year.

You can find these wettish features in every state of America, true, but not in the same astounding volume as here in Alaska. Because of frozen ground that does not allow water to drain away, Alaska is a wet state. There are no desert hikes here. Water is easy to find almost everywhere.

Someone once calculated this: In the 800 miles the pipeline traverses across Alaska, that relatively straight line intersects 800 rivers and streams. I did not count each one, but that number feels accurate to me.

There, where I stopped to add water to my bottle, I noticed something that holds true even when I am not hiking across America's largest state: Seeing wild creatures, little and big, makes me happy. It's why I spill sunflower seeds on a platform feeder at home.

Scientists led by Joel Methorst at the German Center for Integrative Biodiversity Research in Frankfurt authored a recent study. They concluded "bird species richness is positively associated with life-satisfaction across Europe."

The scientists based their results in part on a quality-of-life survey completed by more than 40,000 Europeans from 34 different countries. People valued birds as much as they might finding a $100 bill on the sidewalk. For a couple reasons:

1. Birds are cool, with their ability to fly and their colorful feathers and their occasional cuteness, and 2. Birds perch in trees, hop across sand, and float in water, all in landscapes that — even if man-altered for centuries — tend to soothe the human soul.

If I needed to, I could have figured the chickadee-happiness connection using the formula devised by the scientists:

$$LS_{ijr} = \alpha + \beta \ln(Y_i) + \gamma X_i + \delta G_j + \varphi \ln D_j$$

But I don't need to. I think that study confirmed something intuitive: Humans are happier outside of four walls. A lot of that has to do with the feeling we get when we hear the first robin song of spring, a universal experience for everyone on the continent of North America.

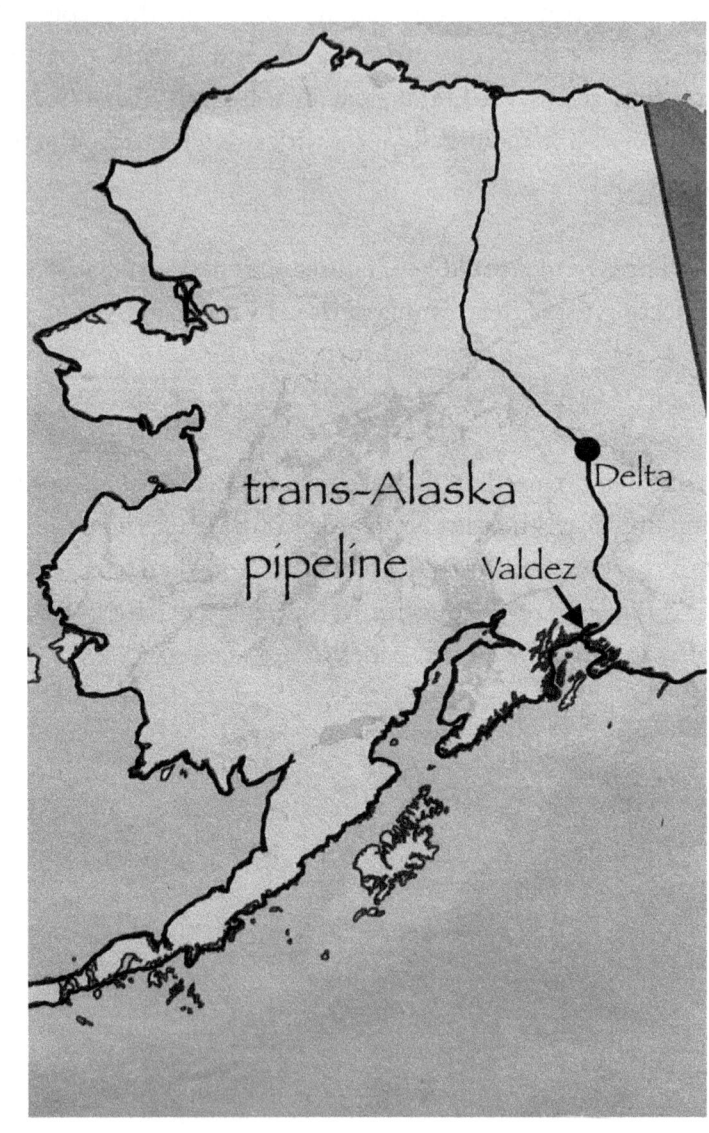

9

Into the boreal forest

Night 33, June 1

For the first time on the trip, the tent is too warm. Cora pants doggy breaths while resting on her foam pad. Stripped to my underwear, my skin sticky, I lay atop my down sleeping bag. The tent fills with trail musk — a mixture of cow pasture and dog biscuit.

The nylon tent flap, which I closed every night up in the Alaska Range to delay the departure of warm air molecules, is open to the sky. The inner screen, though, is still zipped shut. Few mosquitoes have emerged yet because the summer warmth has just begun, but one pair of whining wings is too many.

I am resting on crunchy lichen within Fort Greely, a military base in interior Alaska. Near me, missiles are buried in the ground, ready to fly in defense of the country. The workings of the Army post — home to part of the National Missile Defense System — are a mystery to me. But on a nearby access road trucks and cars move in and out like busy ants.

From inside the tent, I just heard a recording of *Taps*, played on trumpet. That unexpected, mournful tune, wafting over the black spruce forest and broadcast into the wild by someone on base, has me tearing up. It's the somber nature of the tune I guess, and the feeling of time flying by.

The melody takes me back 35 years. Back then, I was an airman trying to catch grayling on a slough outside Eielson Air Force Base. Then, too, *Taps* had taken me by surprise; it seemed to come from Heaven while I was standing at 10 p.m. with rubber boots in a stream. I remember standing at attention in the water, raising my right hand to the brim to my baseball cap. I wasn't one of those ate-up military guys (I only lasted my minimum hitch of four years), but *Taps* had that unexpected punch.

Back then, as a 19-year-old, I found my fish-seeking-self surrounded by a new, fairy forest. The trees were a shorter than those I grew up with in New York, but seemed wilder. In the lowlands were black spruce not much taller than me, crooked as an old man, most of them a century old. Balsam poplars with big, sticky leaves that rattled in the wind. Aspens with leaves on stems as thin as sewing thread whispered their

SUMMER OF GRAVEL AND STEEL

Latin name, *Populus tremuloides*. To me, the boreal forest was somehow both foreign and familiar.

Though Cora and I have been dabbling in the boreal forest all the way, our path will soon enter the heart of the great wooded wetland of Alaska, one that sweeps across the northern part of the continent, all the way east to Nova Scotia. The forest jumps oceans to encircle the top of the globe, straddling the Arctic Circle through Scandinavia and Russia.

Boreal is Latin for "northern," and that ecosystem of trees and rivers and swamplands does not exist in the Lower 48 states, but it's *the* forest in Alaska, at least away from the coasts. Wherever there are trees in the wide sash draped over the middle of the state, they are plucky spruce and dressed-up paper birch.

* * *

A few days later, nearing the Tanana River, Cora and I walked into the boreal forest proper. This northern land where I have slept for the past 30 years is the realm of two evergreen trees — the twisted black spruce and its brethren, the taller, spear-like white spruce, the stems of which swell thicker than a man, stout enough to enable their use in log-cabin construction. The forest is also salted with paper birches and balsam poplars, aspen, tamaracks and 33 species of willows.

I was born far from this green unibrow sprouting across

North America. My factory hometown in New York — Hudson Falls — is located in the great mixed hardwood forest: Beech and shagbark hickory and sugar maple. The nearest boreal forest is about 250 miles north, in Quebec.

Over decades, a human's familiarity with any landscape gets stifling. A friend of mine here who is of the same vintage and who also grew up in New York craves the sunshine of the west-Texas desert during the subarctic winter. He talks of the lack of surprises in the calls of chickadees, the sting of the cold, the squish of reindeer moss, and the tang of spruce needle in his oatmeal. He likes visiting different ecosystems, all of which are closer to the equator, and all of which are much richer in creatures than our hungry home in Fairbanks. I feel his pain, especially November through February.

* * *

This Alaska zone of trees and lowland bogs is so big it will take me and Cora two months to traverse it. From the north side of the Alaska Range, the boreal forest spills over Interior Alaska, into which you could paste Iowa. The forest then reaches its fingers up warmish drainages of the Brooks Range. North of those mountains, ecologists have figured the climate is too cold for trees to germinate into delicate saplings and still have enough frostless days to harden for winter, so they don't grow there. Aside from some scattered poplar groves, the Arctic features nothing you would call a forest.

Since the mid 1980s, when I chose to drive from the Empire State to Alaska in a four-cylinder Ford pickup, this has been the landscape of my life: A broad, hilly surface with Labrador tea smelling of turpentine, ripe blueberries that fire your salivary glands, the springtime drum of far-off grouse thumping the air. Big, wide rivers like the Tanana and the Yukon, carrying gritty water hundreds of miles in low-gradient flows with almost no constricted rapids. Millions of large animals, spread far enough apart to be — to the disbelief of newcomers to the state like Airman First Class Ned Rozell — very hard to see.

The wolf is one of these. Overlay the map of the boreal forest on one of nighttime North America lights and you notice a few things: One, all of the large, glowing cities where we leave lights on all night are south of the boreal forest, and two, if you lay the range of the gray wolf over the continent's electrical grids, wolves fit nicely into the dark evenings of the boreal forest.

Wolves have trotted away from the light. They are one of several species who seem mutually exclusive from humans — except for the few people living in the extreme far north, who are willing to turn off the lamp at night.

Study the map of this immense northern forest and you see only a few outposts that might be called cities: Anchorage, Fairbanks, Whitehorse, Edmonton, Yellowknife. We are spread far apart, but we have this in common: We have a few lug nuts loose, choosing to live in the cold, dark and buggy.

* * *

We people of the boreal forest also share the real danger of burning up. The evergreens surrounding us and their dry, dead branches that finger to the ground are designed to ignite and efficiently spread flames.

In my time in Alaska, curved branches of spruce trees have touched my house in every place I have lived. I once helped my 85-year-old neighbor toss split rounds of spruce into his woodshed. I mentioned to him that many trees seemed to be within falling distance of our houses. I wondered what we might do if they caught on fire.

"We rely on suppression," Dave said, noting that firefighters might be able to save our homes — just north of the University of Alaska Fairbanks — due to our proximity to town.

I looked around at the green-and-brown thickness and kept my doubts to myself. I remembered working on a wildlands firefighting crew with a man named Rodney. He chose that line of work for a reason, which became apparent when he once whispered to me.

"Watch this."

Rodney flicked his yellow Bic lighter. Grinning to show the gap between his two front teeth, he touched the flame to a branch of a black spruce tree. The fire licked the needles and climbed upward with surprising speed. In 10 minutes, the tree was a torch, popping and woofing, forcing us to back away from its squealing heat and blue-black smoke.

I was speechless as Rodney grinned at his work. Other firefighters hustled over to chainsaw down the burning tree. It fell with a shower of sparks. We smothered the flames by beating on them with the heads of our Pulaskis — hand-tools with dueling heads of ax blade and hoe.

I was too stunned to rat out Rodney, but kept an eye on him. I didn't see him set any more trees on fire, but his lighter confirmed the words a forester used to describe black spruce: Gasoline on a stick.

Burning is what the boreal forest does. Black spruce have semi-serotinous cones. They open when dry and after being scorched, when tan cones spread open shed seeds, which helicopter to the blackened ground.

Wildfires leave behind fertile opportunity for other trees, like willows, the buds and leaves of which are the preferred food of moose. A few seeds that waft into burned areas germinate, take root and continue the cycle. Spruce, tolerant to growing in the shade of other trees, are usually the last trees standing, before they burn again. Though we freak out when they approach our stuff, fires are a natural, essential function of this ecosystem.

* * *

In 2016, we in Fairbanks got a preview of what might happen to our town while watching news coverage of the burning of Fort McMurray, Alberta. That northern-Canada town is about where the buckle might be on the boreal forest's belt across North America. At the time, Fort McMurray was home to about 100,000 people, similar in size to Fairbanks.

There, during an unusually hot, dry May, the forest caught fire and spread so quickly that overwhelmed firefighters could only watch as the wind spread flames from spruce to spruce to spruce.

It was the costliest disaster in Canada's history — about 10 billion U.S. dollars — in the fighting and the damage done. The 1.5-million-acre fire consumed 3,000 homes like mine and my neighbor's, about one-fifth the town's total.

Like Fort McMurray, many if not most Fairbanks homes are snuggled into the boreal forest. Several times in recent years wildfires have burned close enough to Fairbanks to cause some homeowners to evacuate. State firefighters have cleared "fuel breaks" north of town but officials remind us that those narrow strips of cleared land do not guarantee our safety.

SUMMER OF GRAVEL AND STEEL

A few years after my neighbor died, a wildfire did indeed spark up on the 1,000 university acres adjacent to our homes. Dave was right: A team of wildland firefighters and water-tanker flyers extinguished the flames in a 24-hour battle. The ground-based, Pulaski-and-chainsaw wielding Angels with Dirty Faces staged out of our driveway until the fire was out. Three acres of charred forest were left behind.

In the 20 years between the times of my first pipeline hike and my second, 31 million acres of Alaska burned in wildfires, most ignited by lightning. That's a black scar the size of Pennsylvania.

I remember the summer of 2004, when a Vermont-size patch of Alaska burned in a few months. Valleys and hillsides were on fire north, south, east and west of Fairbanks. No matter which way the wind blew, we marinated in orange air. When Kristen and I punched out of the smoke on a 737 to attend her sister's wedding in Colorado, we noticed at her sister's house that all our clothes smelled like cigarettes.

The nostalgic tang of spruce smoke is part of the summer experience here. I like that smell for how it reminds me of past summers — until I can't escape it. Then I feel trapped, like I do at minus 40, when the frigid air and ice fog feel as suffocating as a thick blanket of smoke.

Scientists have predicted that in future warmer summers brought on by climate change, more of the boreal forest will burn. We are probably living through a long transformation from boreal forest to open aspen grassland.

* * *

Because this place is so flammable, I chose not to make campfires in either 1997 or 2017.

I had a few other reasons for abstaining during my hikes. One: I probably shouldn't start a fire near the pipeline. Two: I was so tired from walking at the end of each day I never thought a campfire would be worth the significant effort of gathering dry wood. Three: Campfires are the cave-man's cellphone, into which I stare all night. I would rather notice other things, like the breeze through the leaves, or the robin hopping from branch to branch with tufts of nesting material clamped in her beak.

* * *

Cora and I found little water to drink until we walked upon the south bank of the Tanana River.

Reaching the big river after nearly 300 miles of steps, Cora and I sat in the shade of a poplar and watched the water flow by. The Tanana was flat and tan, dimpled by eddies. Swallows darted over its surface, their calls sounding like they were chewing rubber bands.

My feet cooled after I released them from my hiking boots and exposed them to air. Leaves, uncurled from aspen buds, soaked up energy, each a dime-size solar panel. I could almost hear them stretching. At the start of June, it felt like summer.

As my feet enjoyed their freedom, I pondered the Tanana. It's pronounced Ta-na-naw, rather than — as I assumed as a young airman — the last name of former Major League Baseball pitcher Frank Tanana, which rhymes with banana.

A Poor Man's Yukon, the Tanana has all the adventure but easier logistics — it flows near paved highways for more than half its length.

Over the years, I have in sections canoed the Tanana's entire 550 twisting miles, from where the glacial Nabesna River collides with the Chisana River, near the village of Northway, down to the village of Tanana, where the river dumps into the mile-wide Yukon.

One big difference between the Tanana and the much-larger Yukon: Mud. The shores of the Tanana are often aproned with sticky brown silt, which reaches deep into the trees. On several legs of my float of the Tanana, it was hard to find a chip of gravel, or a carpet of pebbles to keep my tent floor clean. On the Tanana, you make peace with grit. Sections of the Yukon I have floated have featured gravel-ringed islands, superb for camping.

The Tanana twists sometimes in a single silty channel and other times is split in wide, flat braids flowing through what seems like a million forested islands. It is totally boreal, always flowing through spruce forest.

On one of my journeys, from Nenana to the village of Tanana (where the river enters the Yukon), my friend noticed white-winged crossbills, handsome red-and-yellow birds that use their fingers-crossed beaks to tear spruce seeds from cones. Like the other birds we saw on our journey, they were familiar to us.

"We've traveled a week and it's like we're still in our backyard," he said as we absorbed the pokes of mosquitoes at an abandoned village site. I could not argue with his logic, and got a sense of how large the boreal forest really is.

The Tanana of the boreal forest may not offer a journey through changing ecosystems like a hike up a desert mountain, but it is now my home river, replacing the Hudson of my youth. To me, the Tanana fits me like my favorite Ikea chair, the one in which Cora sits on my lap as I sip my first cup of coffee, often in darkness.

The Tanana does not get much love in Alaska river guidebooks, probably because the water is brownish, carrying the powder of mountains ground down by glaciers. Most people choose to float clearwater rivers that are better for drinking, and with a bit more splash to them.

The appeal of the Tanana River to me is similar to my decades of boreal-forest existence in Fairbanks: Living here is not going to get your heart beating too fast.

But there is a *space* to this place that I become acutely aware of when on a crowded trail in Colorado. That freedom doesn't come free — the air hurts in winter. In summer, the bugs make you sometimes wish for winter.

But living here gives you an appreciation for things, like the sun: The solstices never pass without a bonfire, or at least nod of recognition.

After a while, the boreal forest wrapped around me like the sweet smell of spruce smoke. As many have noted, once you live here for a long while, you are unfit to live anywhere else.

10

Walking home

I suddenly had wings on my feet. Kristen and Anna brought them to me. They visited me in Delta Junction as I stayed overnight on the south bank of the Tanana River.

Because summer had arrived in early June, I no longer needed the warmth provided by the synthetic padding of my hiking boots, in which I had taken a few hundred thousand steps from Valdez to middle Alaska. To replace those clunkers, my thoughtful wife and daughter handed me a box of new running shoes, La Sportiva Mutants.

Sneakers, we used to call them. To me, they felt weightless, wrapped like slippers around my ankle-high runner's socks and my toes, which were hard-and-sharp as a raven's from their mileage inside the boots.

With Cora on a leash, I hiked over the Richardson

SUMMER OF GRAVEL AND STEEL

Highway bridge to cross the Tanana. To our left, the big river crashed silently into the Delta River, which spilled north from Alaska Range peaks. There, the engorged Tanana hangs a right around a rock bluff and heads, wide and brown and smelling like dirt, toward Fairbanks.

Once across the bridge and after some wet ditch-walking on the side of the road, Cora and I re-gained the pipeline and the gravel path next to it.

On that day, we clipped off a new record — 14 miles.

Why? My new wings. I was no longer swinging double the weight on my feet with each crunchy step on the pit-run gravel. Trail angel Kristen had also exchanged my winter sleeping bag for a featherweight summer down bag rated to 45 degrees. She had, too, loaned me her sleeping pad, so small you could shove it inside a water bottle, half the weight of the one I had been carrying.

On the north side of the Tanana River, Cora and I were then less than 100 miles from our home in Fairbanks. When I reached my house, I would not yet be halfway on the journey from Valdez to Prudhoe Bay.

In early June, there was still plenty of summer left. I was more than a week ahead of my progress at the same date 20 years before.

Was I really faster as a 54-year-old than when I was 34? My times in running races had been riding the downslope of the life arc, but my mindset was a bit different than it was in 1997.

In the 20 years between pipe hikes, I had signed up for a lot of events like the White Mountains 100-miler and the Alaska Mountain Wilderness Classic Ski Race. Those races changed my view of what's possible. Longer days were doable, so long as I was efficient in camping, didn't waste time and got my rest. On this trip I was averaging about 10 miles each day, compared to six on the first trip. Those extra miles add up.

* * *

Night 39, Gold Run Creek, June 7th

Cora and I are way far from anything. The pipeline here has bent 12 miles from the Richardson Highway. Besides me and Cora, there are insects and gray-cheeked thrushes and Swainson's thrushes and ruby-crowned kinglets.

And mosquitoes. Tonight was the first time I spritzed my neck with insect repellant, on night 39 of the trip across Alaska. Acceptable.

We are in John Haines country, in the back of his expansive trapline range that fronted the Richardson Highway. John was a poet and essayist who captured Interior Alaska as well as anyone.

The late storyteller once sent me a letter in response to a newspaper column I wrote about the shipping network that allows Alaskans to eat fresh broccoli in midwinter.

In his handwritten note, Haines reflected on life in Alaska decades ago; how he ate honey from northern bees, muscle-and-organ meat from moose he shot and potatoes he dug from his garden. And he remembered the lean times, when he did not find a moose and he dropped the lean carcasses of snowshoe hares into his steaming pot.

John told me he was not thrilled with the prospect and then reality of this pipe running through the back forty of his wilderness. But it came, it is here, and the pipeline always has been part of my Alaska.

Gold Run Creek seems wild, even though I'm leaning against a vertical support member made of steel imported from Japan. It's quiet enough for me, anyway. I get to see songbirds up close all day and, now through 40 days, have yet to see a bear.

That's a close-up siting I can do without. John Haines once wrote in his spare, slow cadence of shooting at a grizzly that charged him from a small creek bed not far from here. He perhaps wounded it with a shot from his rifle. He didn't know if his bullet found the mark, and he found no spots of blood. The bear retreated to the alders crowding the creek.

Haines crossed the water with his trusted sled dog and continued on to one of his trapline cabins. A few days later, he needed to again transit the creek to make it back to his home on the Richardson Highway.

"If that bear was still somewhere in that dense green cover, nursing its hurt and its temper, waiting for revenge, it would have its chance," he wrote in his memoir The Stars, the Snow, the Fire (in my backpack for this section).

On his return trip through the creek, Haines did not see the bear again. Neither did we, so far.

Night 40, Salcha River, June 8th

Or is this Maui?

We've got 78 degrees at 10 p.m., Cora and I, on a massive white cobble bar free of bugs! The good life.

It was that way most of the day as the temperature climbed above 80. There was plenty of permafrost-suspended water to keep Cora alive and happy in the heat, maybe the warmest air we will feel this summer. Though her dash for a red fox here on the floodplain was ill-advised. She still pants from that sprint.

It was so steamy today that the moose were seeking watery

refuge. Confused as what the dark shape was at first, I saw a head with its enormous ears bobbing just above the surface in one trailside lake. The moose's giant, bulbous body was hidden underwater.

The moose scented us and looked my way. I waved my hat so it could detect a bit of movement. The huge cow noticed and slogged out of buoyancy, dripping gallons from the pond into the hot muskeg. Sorry about that.

It is nice to be out here on the rocks away from the mosquitoes, which stick to the brush and prefer it cooler. Feels like a river trip. I've blown up the small yellow Scout packraft that folks at the Alpacka company have loaned me. Ready to deploy tomorrow. I'll need it to cross the river, which the pipeline is buried beneath.

One of the best nights of the trip. Wish someone was here to share it!

Night 42, Eielson Air Force Base, June 10th

Camping in a spruce hollow by French Creek Bridge No. 1 with Bobby Gillis.

Bobby, a geologist, is a year younger than me. He has curly hair, glasses, a moustache and goatee beard. He plays hockey, drums, and — unlike any of our mutual friends — likes to ride dirt bikes of the motorized kind.

I have a nickname for him: Nails. Once, when I noticed he was not carrying a packraft to cross a major river slicing the country in the middle of a 130-mile race, he explained.

"I'm just going to walk in the water until my feet don't touch bottom anymore. Then I'm going to swim."

One of my favorite Bobby images is of him breaking trail on a 150-mile ski trip that six of us executed from Shishmaref to Nome on the Seward Peninsula. He was clicking along on skis up a mountain, on a trail marked by wooden tripods in that naked country without trees. The pole trail markers, looking like teepee frames, were the only black on a pale white landscape.

Bobby was skiing so fast, even while plowing through drifts of fresh windblown snow for us, that we couldn't keep up. I played Stairway to Heaven in my head, because that's where it looked like he was disappearing as he ratcheted away into the white.

Bobby had texted me on my inReach and said he wanted to come along on Pipeline Hike II. It is nice to have dinner company, as we heat water we scooped from French Creek while we sit on tarred timbers of a bridge.

Bobby is more a listener than a talker. We noticed flickers shrieking amid the spruce as we ate. When we chatted, we contemplated reaching his car at a nearby hamlet called Moose Creek, probably tomorrow.

Cora and Bob

That tomorrow arrived with a pissing rain. I woke in the dampness of my tent with a headache after not sleeping much.

I had trouble sleeping because it never got dark outside;

the tent glowed at all hours. Light always fell on my eyelids, playing havoc with my circadian rhythms. It's a common Alaska summer malady some people cure by sleeping with all their blinds closed, or by pasting aluminum foil over their windows. I made a note in my journal to buy an eye-mask when I got to town.

Bobby and I ate breakfast on the bridge, rainwater dripping from our noses into our oatmeal. We each swallowed half a dozen mosquitoes that stuck to the oatmeal on our spoons.

We were eight miles from Bobby's car, parked at the Moose Creek General Store. From there, it would be about 23 miles to Fairbanks. I had a decision to make.

The pipeline crosses underground through the nearby town of North Pole, traversing the backyards of several landowners. No public easement offers people legal access, so, like 20 years before, I had planned not to walk this section.

I had an alternate route in mind. As I was still carrying my packraft from the Salcha River crossing, I could — from the Moose Creek dam spillway — walk just a mile down to the Tanana River. I could inflate the raft, shove off into the big river, and float to a landing within 4 miles of my home. From there, I would hoof it to my front door.

That plan had a satisfying purity to it: Though not

following the pipe, I could still use my boat to continue my human-powered journey across Alaska.

Later that day, Bobby and I and Cora hiked along to Bobby's car after finding some subdivision gravel roads that led that way. The rain did not let up; more than 1.3 inches would fall on interior Alaska that day. My headache had not faded, not even after drinking three Vias. Shivering already, I thought of sitting in an open packraft in a big river with a wet dog, her pleading eyes upon me for six hours.

"Robert," I said as we were squishing our way toward his car, "I'm whipped. Can you just give me a ride to my house in Fairbanks?"

"Of course," he said. "Anything you want to do."

"I feel like such a wimp," I said.

"It's your trip," Bobby said. "You can make your own rules."

* * *

On June 11, 43 days after I started, I returned to my house just north of the university in Fairbanks.

We had rolled out of the driveway in late April, when snow was the dominant ground cover. I came back to so much *green*

— leaves waving at me like little hands, all exhaling oxygen into the warm air. When I was last home, they were all curled tight in buds, hovering over a brilliant white blanket.

Staring at the boreal jungle from my back deck, I was stunned. It was as if the world viewed from my coffee-drinking chair had changed from winter to summer in just one day.

Cora and I had walked 350 miles in that month-and-a-half since our first snowy steps in Valdez. We had 450 to go, but not before a break.

Bobby had driven me 31 miles from the Moose Creek Store to home. That ride in his car, windows steaming with the moisture of three wet bodies, felt a bit like cheating. But in the context of the trip, it was just a blip, soon to be forgotten. Self-leniency is one of the advantages of growing older, right?

I had some chores at home, the most pressing of which was to fill 11 cardboard boxes with four days of food each. Those would be my food drops for the northern portion of the hike. I would draw a number on each box with a Sharpie.

These tasks I did alone. Kristen was on Alaska's North Slope studying birds. Anna was then in Colorado with her cousins. The house was lonely and quiet.

In addition to packaging my food drops and stacking the numbered boxes in our little guest cabin (for others to deliver to me on the pipe), I partook in activities I love during the unending light of midsummer Fairbanks.

I went with a friend to an Alaska Goldpanners baseball game, my back warmed by the sun as we stood on the top row of the left-field bleachers. I played shortstop for my softball team, the Northern Shrikes (and struck out twice). I played Wiffleball with Ian and Chris, hitting the plastic ball onto the roof of their house. I had dinner at the Pump House restaurant with my friend Andy and his parents, who visited from Massachusetts. I sat on my deck in the sunshine and focused binoculars on one of our birdhouses, watching two adult red-breasted nuthatches shove insects to their nestlings and then fly away with pouches of white poop that resembled diapers.

I spent six days at home, with much of that time devoted to the filling of the boxes that would provide all the dog food and fuel cans and dehydrated dinners and candy bars and bug dope and Vias and toilet paper we would need for the rest of the summer. More than half of the Alaska map was waiting ahead — all we had to do was to unstick ourselves from the glue of town.

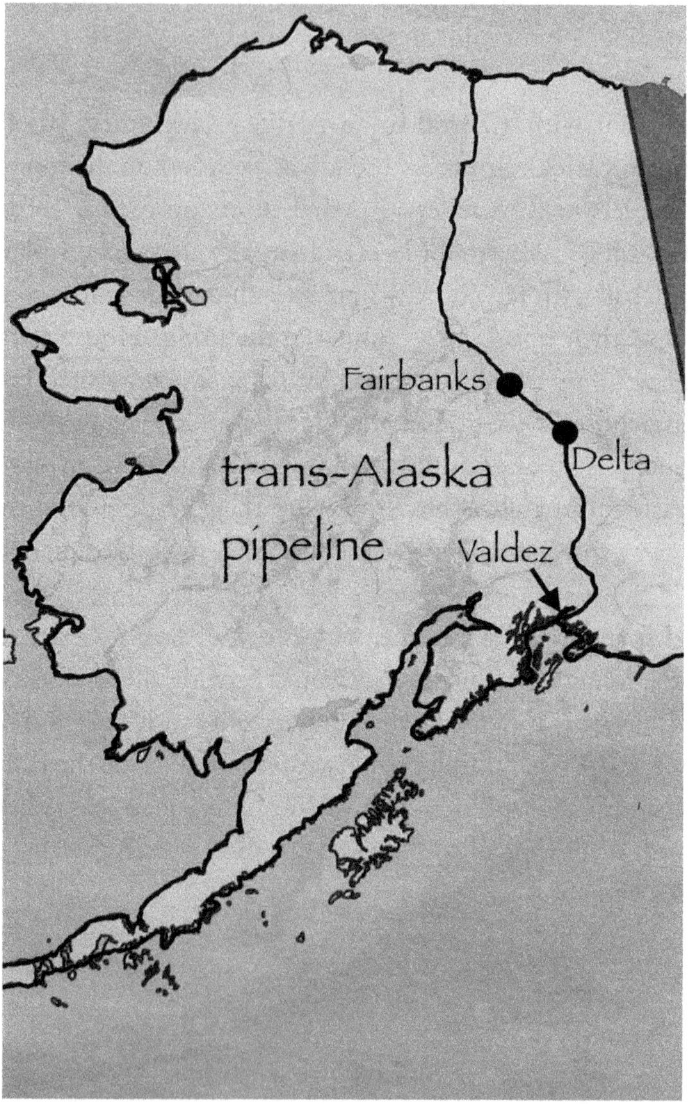
Fairbanks
Delta
trans-Alaska
pipeline
Valdez

11

Cora

As he did 20 years earlier, my friend Andy Sterns gave me a ride to the Pipeline Viewpoint in Fox, Alaska. Balancing on my lap, her claws digging into my quads as she peered ahead through the windshield, was Cora.

On both hikes, I had bypassed the Fairbanks section of the pipe as it snakes through a few dozen miles of private property in town. Trying to get permission from all those landowners would have been exhausting; many would have said no anyway.

My purity had already evaporated with the ride Bobby gave me from Moose Creek to Fairbanks. It was no big deal to bypass the peopled section of the pipe. The wilder, more appealing part of the pathway — the part away from peoples' yards — was just ahead.

Cora, anticipating freedom was near, started whining as Andy drove. After six days in which we both slept in my comfy bed, my dog and I were pulling free from the stickiness of town.

Andy turned his green pickup into the Pipeline Viewpoint, a spot just south of where the pipeline worms beneath the Steese Highway. That pull-off is one of the few sections of the pipe Alyeska officials welcome non-pipeliners. The viewpoint features a few interpretive signs about how men and women built the 800-mile pipe in a few short years, and how engineers routed it above-ground in areas such as that one, where soil frozen since the days of the steppe bison had thus far remained hard as rock.

I pulled my backpack out of the bed of Andy's pickup, then went to his passenger door, where Cora pleaded to me with her brown eyes. I opened the door; she flew like a bird from a cage.

I then hugged the muscle-and-sinew slimness that is Andy — a smiling presence in my life since before the 1997 hike — and stepped onto the Steese Highway. I clipped Cora to her leash, and the loop of that to the waist buckle of my pack. Cora yanked me to the other side of the pavement, the tether as stiff as copper wire.

Following the pipe's underground dive beneath the highway, we again stepped on gravel. There, we launched into our final 450 miles across Alaska.

I felt the joy of swinging my legs in their hip sockets, muscles and joints happily working as they had throughout the trip. My pack, an old friend, clung comfortably from my shoulders and waist. Just ahead, my little black pal, freed from her leash, torpedoed the willows.

Hearing the twigs snap from her exploration, I wondered if my dog detected a few familiar air molecules from the hillside of her birth. There, nearing the community of Fox, we passed within one mile of where she began life on Gold Mine Trail.

* * *

A little more than three years earlier, a neighborhood blue heeler had gone wandering in the Gold Mine Trail neighborhood. He sprung over a chain-link fence to meet Cora's mother, a yellow Labrador retriever. Their union led to a crop of puppies, soon advertised on the internet.

One day around that time, during one of the rare dog-less periods of my life, I noticed my wife and daughter looking online at photos of puppies.

A few months earlier, we had lost Poops, a sweet and insane Lab mix. She had died too young, from a tumor. Right away, we felt the void of not having a creature driving her soft head into your knees every morning, a sock dangling from her mouth.

Though I missed Poops' crashes into the brush as she chased hares and returned with a bloody forehead from sharp branches, I confess to enjoying the break from that daily responsibility. I figure we owe our animals a daily loop, long or short, and have done my best to execute.

The puppy-viewing quickly escalated to its One Sure Outcome: Kristen found a crop of tiny black dogs. The owners would give the dogs away to a good home: First-come, first served with a good-smelling, wriggling body and its feather-soft bulbous belly.

* * *

On a Saturday morning in springtime, we coaxed 7-year-old Anna into the car without telling her the purpose of the drive. We ended up at the home of John and Liuda Eichelberger on Gold Mine Trail. That road, north of Fairbanks, is located in a subdivision with names like Grubstake, Dredgeview and Pennyweight.

John is a volcanologist who had 20 years before led me and others on a trip to the Valley of Ten Thousand Smokes in Alaska's Katmai National Park and Preserve. He impressed me then by hiking our group across the abrasive pumice to the summits of windy peaks where I wished for an ice axe that no one carried. Bearded and jolly, John sang nautical songs from the days of the Yankee whalers while balanced on those windy, knife-edge ridges. He was not the least bit scared.

When Kristen, Anna and I entered he and Liuda's home, the living-room floor squirmed with nine black bodies. Anna's surprised smile, firing her dimples, was worth 15 years of commitment.

Anna chose the smallest dog, identifying Cora by tying a string of blue yarn around its tiny collar. We could come get her in a week, John said, when Cora was further weaned from her mother.

One week later, Cora first demonstrated her outsized voice as John (was that a tear in his eye?) carried the black puppy from her mother and siblings. He handed her gently to Anna, sitting in the back seat.

Cora's wail in the back seat was heartbreaking, the cry of a creature realizing that everything it once knew had changed forever.

I had a flashback to Anna's first day at the university preschool, when I handed her over screaming and flopping like a salmon to a grandmotherly teacher who fired a look that said I should get the hell out of there.

My ears smarting, I wasn't sure what I thought about that puppy. Cora's nose was stubby, like a boxer's. She also had an underbite. I thought a good name for her would have been Muggsy. Anna chose Cora.

Cora would remain the runt among her eight siblings. Even after a year of growth, she did not top 40 pounds. This, we later found, was one of her best attributes, most practical in a canoe, or when she climbed in our friends' laps. And Cora is small enough to sleep with Anna, curling up behind her knees without crowding her.

Cora cultivated her gift of loudness, an explosive bark that is shocking from a body so small. She blasts it when she sees another dog through the window, hears a knock on the door, or partners with Fluffy — the neighbor dog who started the trip with us and Cora's best friend — in attempts to fell a red squirrel from a tree using sound waves.

Poor Fluffy. She has, like us, suffered hearing loss from Cora's excitement. And when Fluffy is wearing her own bark collar, she sometimes gets shocked by Cora's enthusiastic barks. Sorry Fluffs, but that cracks me up every time.

That loudness proves useful in some situations, though, as Cora proved on down the pipeline when a bear was near the tent. That story comes later.

Cora has proven to be an excellent canoe dog. Her

wanderings in the boat are easy to compensate for with a butt scoot. She clings to the gunnels like a housefly, her feet somehow balanced on the narrowest of beams, even when I try to rock her into the river. After a wave knocks her off, or when she jumps out upon seeing a beaver's head cutting through the current, she eventually swims for shore. Or she paddles alongside the boat, where I can grab her collar and use her buoyancy for a quick snatch, hauling her dripping body back in.

* * *

When I was home boxing food drops, my neighbor Audra Carlson offered to give Cora a break from the trail. Audra said Cora could stay in Fairbanks for a while and sleep in a dog bed next to Fluffy.

I pondered that for a few minutes, thinking back to the first 350 miles of hill and valley we had covered since Valdez. Cora had weathered her millions of steps quite well, I thought.

I had carried her backpack in my hand for a few days when the fabric rubbed her raw near a front leg, but we had solved that problem by tightening the straps. She once stepped on a thorn and started favoring a leg, but I pulled the tiny spike out with tweezers. Her limp disappeared.

Though she had no vote in the matter, Cora seemed to

enjoy the outdoor life. I told Audra thanks, but Cora would go the distance.

At night on the trail, Cora pressed against my lower back as I sat next to mountain streams and typed stories into my iPad. In the tent, she slept next to me on her own foam pad. Some chilly nights she woke me up with a lick to the face. I would peel back the neck of my sleeping bag; she would slither in. Somehow, in the skinny tip of the bag way down at my feet, she turned around and curled up to sleep.

Mummied at my toes was not the protective position in which I preferred Cora. Each night, after I stuffed in earplugs, I depended on her ears to alert me to the scuff of gravel or the snap of a twig outside the tent. But sometimes she was deep within the bag, twitching in dreamland.

Cora minds my voice commands. *Come* brings her running to me when we see a bull moose ahead. Some dogs, like one that joined us early in the trip (not Fluffy), can't resist the call. That dog brought an angry momma moose back to Anna and her friend, who escaped by running up a little hill. Cora seems to know not to mess with big creatures.

Cora barked at the only porcupine she saw on the first half of the trip, in an alder near Valdez, but she did not go after it. I gave her a few strokes and a *good girl* after that one. When dogs intersect with porcupines, the imbedded quills in the dog's face and the dog's resistance to having them yanked are a true show-stopper, even if one is lucky enough to be near a town, or have a pair of pliers along (which I did).

Cora has short hair. She loves the water. Her joyous dives and swims keep her clean and cool on the steamy days.

She suffers in the bugs, but when it gets real I squirt Deet on the back of my hand and apply it to the bridge of her nose, being careful to avoid her eyes, nostrils and mouth. She takes the dosage with stoicism.

Cora's one flaw, nullified on the solitary pipeline trip, is her occasional chippiness upon meeting other dogs. Upon greeting another dog on the trails, her hackles raise into a Mohawk. She becomes a little punk, the undersized hockey player ready to throw the first punch.

At three years old during the hike, Cora was at the peak of life. I saw an energy contrast between she and Jane, who accompanied me 20 years earlier. Jane was 10 years old on that trip.

Cora recovered faster than Jane, who sometimes shied from the pack in the morning. Cora wagged her tail and gleefully mauled nearby sticks when she saw me approaching with her pack. She knew it meant we would be moving soon.

Cora became jacked like Giancarlo Stanton during the walk from Valdez to Fairbanks. Her legs were four turkey drumsticks. I guess that's what happens when you bust through the brush all day, every day, with the resistance of a doggy backpack full of food.

In midsummer's unending light, on I went away from Andy's truck with my agreeable companion. Half of Alaska spread before us in the blue hills, waiting for our footsteps.

I thought back to my neighbor's offer to keep Cora for a while as I continued ahead. It became evident to me that sharing the entire trip with my little buddy was one element of purity I needed to keep clean. I truly did not want to go on without Cora; it was her trip as much as mine.

She confirmed this to me as Andy was driving back toward his home. She busted out from the alder leaves and confronted me panting, her mouth wide open with a doggy smile. All the voles in northern Alaska were in front of her, and she knew it.

12

The real deal

In its 800 miles of being strung over Alaska, the pipeline experiences some of its most extreme ups and downs just north of Fairbanks.

As I dropped to Treasure Creek, crossed a bridge over a trickle of water, and then ascended out of the valley, I found myself kicking footholds in the gravel road beside the pipe.

The engineers who decided on the pipeline route pretty much drew a straight line for the 100 miles from Fairbanks to the Yukon River. Some of the hills are so steep I can't imagine driving a truck up (or down) the bobsled chutes that pass as access roads for servicing the pipeline

I wondered if even if my bicycling friends who have ridden across the state in winter could remain in the saddle climbing these hills. Seems like hike-a-bike country to me.

Cora did not mind the grade changes, but I found myself taking more snack breaks on the hills, sitting my butt on a rock and looking back from whence we came. Breezes kept mosquitoes away and evaporated my sweat as I squinted back on the silver path, satisfied with how our little legs had carried us from there to here.

At least my pack was a few ounces lighter. While in Fairbanks, I rid myself of a useless item: my new cellphone. I had carried it hoping to generate a wifi signal that would enable me to send my weekly columns back to the Geophysical Institute. And to sometimes catch John Sterling's play-by-play from his booth at Yankee Stadium as I boiled water for dinner.

The wifi plan was a good concept that almost never worked. Not enough cell towers existed between Valdez to Fairbanks to activate my little glass-and-aluminum-and-silicon rectangle. It was dead weight most of the time.

While back in Fairbanks, I deactivated the monthly plan. Friends had told me the phone would not work a few days north of Fairbanks anyhow, except possibly in a small node surrounding the truck-stop in Coldfoot, about a month ahead. Not enough reason to keep carrying it. So I threw the phone into a dumpster.

That action, though wasteful and not the proper disposal method, felt so right.

I never liked that hyperactive wafer buzzing at me with Notifications about Important Things. The poor little thing kept begging for me to react.

Ubiquitous for the last few decades, the cellphone has altered human posture, bringing to mind a Simon and Garfunkel song:

And the people bowed and prayed, to the neon god they made.

I have chosen not to carry one — except for that first stretch of pipe — because I don't want to tilt my vertebrae in that position, and because I already pay for a monthly subscription to the satellite texter in my pocket. There are many times I feel left out (ever try to find a payphone?). But I have found saying "I'll meet you at Bun on the Run at one" to still be effective.

As a dad, husband, sibling, employee, and friend, my lack of a cellphone is a selfish decision. But communication by speech, email, work phone, home landline, or just stopping by manages to happen. And I'm not texting while driving toward you.

I am not opposed to technology and its leaps forward. I have bought heavy into plastic bags, the internet, and pocket cameras that floated the Great Circle Route in brown boxes aboard container ships spewing bunker-fuel exhaust onto the green hills of Aleutian Islands. But I really don't want to spend so many of my remaining minutes squinting at a little screen. Because I know I would.

With my digital-drug-of-choice satellite texter known as an inReach, I had set up a meeting near where the pipeline passes Wickersham Dome north of Fairbanks.

Jay Cable, one of those friends who has ridden the thousand-mile Iditarod Trail on his fatbike in winter, had asked me if he could join me as soon as he heard I was again hiking the pipeline.

Jay is a born Alaskan, having grown up in the Wasilla area. With a tight-end's build and an unusual cheeriness when inhaling a cloud of mosquitoes, he is a Gear Person. Before the hike, rather than do my own research, I saved time by asking Jay what I should buy.

He turned me on to the ULA one-compartment backpack made in Utah that feels like nothing on my back. It is one-quarter of the weight of the external-frame moose-freighter pack I hauled across Alaska 20 years ago.

In the hills north of Fairbanks, up ahead, I saw two dogs approaching. The tan-and-gray smiling husky wore a backpack; the other, white as snow, had no pack due to his senior status. Behind them walked four people; two big, two little. The Cable-Frescos were joining me for a few days.

Along with Jay Cable came his wife Nancy Fresco, who was carrying a listing, oversize backpack stuffed with parenting essentials, and Molly and Lizzie, 11-year old twins who had grown up thinking four-year-olds hiking the Chilkoot Trail was a normal occurrence.

I felt a pang that Anna — who first met the twins at a Halloween party when the three still pooped in diapers — was still in Colorado with her aunt and cousins, and not with me on the trail.

Molly has her Dad's friendly big eyes and his love for outdoor adventure. She gravitates towards dogs and is quick to wrap her arms around a friend whose boot has gotten stuck in the mud.

Lizzie's Elizabethan face resonates with a royal inner peace as she smiles quietly at some notion. I always wonder what she is thinking. Lizzie was growing into more a mini-Nancy each day, becoming both more intellectual and muscular.

But they were still All Kid. Immediately, the girls upped the fun level of the walking: They wove necklaces of dandelions for the dogs and me, raced sticks down creeks we stopped at for water, and leashed spruce sawyers with nooses made of grass, walking the insects around like dogs before freeing them to buzz away.

SUMMER OF GRAVEL AND STEEL

Lizzie, left, and Molly

We made camp at Aggie Creek, a tumble of clear water with a squadron of dragonflies cutting the air overhead.

When an afternoon thunderstorm pulled the heat from the air, the dragonflies disappeared. Mosquitoes replaced them, in numbers that drove us to our tents after a hurried dinner. The voracious females came in such numbers that even the optimist's optimist Jay mentioned how bad they were.

In the tent next door, I heard him being Goofy Dad with his daughters. It's a role that suits him; he has always struck me as someone who seems more comfortable with kids than adults. I remembered Anna's laugh when Jay played mother bird on the 10-mile trail to Tolovana Hot Springs, dangling gummy worms for the girls to eat and coaxing the trio of tiny bodies farther down the trail.

We joined Jay and Nancy and the girls on four of their yearly trips to the hot springs. On those fall journeys, Nancy — powerful as a rugby player — carried 15 pounds of food into the cabin along with her other gear. She somehow also fit Tootsie Pops and Sweet Tarts she promised to the girls upon reaching the next mile marker.

"Our trips are not dentist-approved," she said, "But they are kid-approved."

The junk-food she carried was not part of the Cable-Frescos everyday routine. Upon meeting them at the university, I often noticed the twins munching on a Fairbanks-grown carrot or last night's steamed cauliflower pulled from a recycled bread bag.

Nancy's super-sweet motivational trail techniques worked. She almost never spoke to the girls, ours included, with a serious adult tone, instead using a sing-song voice. She anticipated the girls' need for a snack or a drink before it became a whine. And somehow those tiny bodies covered those 10 hilly miles that have kicked many an adult ass.

I mentioned to Nancy how impressive her attentiveness and tolerance seemed to a dad short on patience when his wife was away on fieldwork.

"I remember being their age and what it was like," she said of the girls as they were marching along in wet sneakers at age 5.

I tried to think back to when I was five. My 54-year-old brainscreen showed only static. Nancy's enlightened consideration of these small humans and Jay's extended childhood are two of the best examples of parenting I have seen up close. Much better than the scrambling I have done to repair the me-first damage of forgetting the stuffed puppy at home, or neglecting to pack the Goldfish.

Nancy wrote a book: *Go Play Outside! Tips, Tricks and Tales from the Trails.* In it, she summed her philosophy of treading lightly, using the human engine that came as standard equipment, and how parenting can be a part of living the way you want to live.

This passage struck me as true to the intentional way Nancy's family lives:

"I want to get places, using simple forms of transportation that don't burn any gas, don't break down, and don't prevent me from seeing and smelling the details of the world along the way. Commuting as exercise. Shopping as exercise. Errands as exercise.

"I want to roam and explore, feel the wind in my face, and find solace in solitude and wonder in wilderness. Hiking and skiing and canoeing as exercise.

"I want to use my body as a tool — to lug our water, build an outhouse, and haul the wood pellets that heat our house. Life as exercise.

"I want to challenge myself, pushing the limits of how far, how fast, how long I can go. I want to find the space inside myself that comes from movement and exhaustion. Adventure and racing as exercise.

"Last but not least, I want to play, chasing my kids and lurching them skyward, rolling and tumbling, laughing with movement and blatant bodily fun. Joy as exercise.

"I want these things and I revel in them. Having children has encouraged me more than it has hampered me."

Ink is cheap, right? It's easy to write about the person we think (or hope) we are. And to post on Facebook the shots of your daughters biking with you 45 miles to a campsite.

But from the first time I met Nancy — as she was pushing a wheeled double-baby-carrier through slush to the university's preschool — she has struck me as the most human-powered human I know.

So many days, there she is, leaning a bit to the left as she jogs slowly from her job as a climate specialist at the university to her home a few miles away. Snow, sleet, rain, 30-below; she is out there, sweat evaporating from her to help moisten the Fairbanks air.

The Fresco-Cables live in a shared community in a boreal swamp, a few miles from campus. Four other families are scattered in those woods; each own a tiny cabin with no running water. After parking your car at the end of a rutted gravel road, you get to their cabin by walking a long path of mismatched wood planking. It takes a few minutes, longer if you are pulling a water jug on a sled.

Their community includes a larger frame-lumber building where they gather for shared meals. Each family takes turns cooking dinner on weeknights. Before meals, those who gather at the table pause for a moment of silence. Community members attend or opt out, eating in their own cabins or somewhere else.

The communal building, which has running water, includes a shower, washer and dryer, and a sitting room with couches and chairs for reading or kid's playtime. It's like having an extended family, one you can take or leave depending on your mood or the busy-ness of your life.

Their community has a name: Tamarack Knoll. Unlike life in a commune, the Cable-Frescos have their own jobs, private lives, and their own living space. "Co-housing" is a way to describe it. Intentional living is another, one that seems to better describe Nancy and Jay. Most people live differently than they do, but Nancy and Jay make their way seem as natural as the black spruce forest surrounding their cabin.

I had screwed up, miscalculating my rations so that I was one dinner short on the stretch with Jay and Nancy and the girls. As we packed the tents for an escape from Aggie Creek's mosquitoes, Jay saw me fretting at the contents of my food bag.

"Are you missing some food?" he said. "We've got more than we need."

SUMMER OF GRAVEL AND STEEL

He handed me snacks and Ramen and more, enough calories to get me to my next resupply, one day away.

When Jay first approached me about coming along for part of my pipeline hike, I was surprised he was interested.

As I suspected with Luc Mehl, I thought the pipeline trip would be a bit tame for Jay. I had watched him, 10 years younger than me, progress from hauling his kids on cabin trips around Fairbanks to nailing wilderness races like the Alaska Mountain Wilderness Classic. On those, he hikes like a bull moose through thick brush, inflates his packraft for a free ride down rivers, then gets out with a wet butt and walks again. He has won the race with his partner Tom Moran. To finish first is a real Alaska badge of honor.

But in walking with me, Jay spoke more words than I had ever heard him say. He was having a good time. He seemed quite happy just to be in the open air, no matter that our path was more road than trail, and oil gurgled southward through a steel tube that was steps away from us.

I was also surprised that Nancy and the twins had wanted to come along, but I was glad they had. Due to the everchanging dynamic of little girls, we had passed the period when the girls and Anna hung out so much we called them the Triplets.

During this trip, I had moments with each of the girls walking beside me. I asked them about their dogs and the school year just passed. They both diverged into subjects of their interest. It was fun to listen to these little humans I had known.

The girls were quiet sometimes as they walked, but their complaints were few. I had admired their different little personalities since they were tiny, and had just adored them because they were so cute and had features that favored either Nancy or Jay. Now that they were stretching out in height and interests, I was impressed but not surprised at the young ladies they were becoming. Insightful, entertaining, and considerate.

At the Tatalina River, we neared where Jay had parked his adventure truck before he had biked back to Wickersham Dome to begin the hike. The girls giggled as they waded across the ankle-deep river in their Crocs, which had dangled from their backpacks for a few days before becoming useful. Stepping across what was perhaps the 400th of 800 waterways that cut the path, I felt a familiar pang, knowing the Cable-Frescos would soon be leaving me.

We hiked out to the Elliott Highway where Jay had parked the truck. When we reached the truck, Jay opened its camper and handed me a bag of barbequed potato chips, and a giant bottle of Diet Coke. I tore open the bag.

Not long ago, I sat in a meeting room in a new building at the university. There, telling us the plight of the natural world and our role in destroying it, was a white-haired man with the Union of Concerned Scientists.

I sat there in the audience, mentally calculating the carbon the white-haired man had helped release by flying up to Fairbanks from Washington, D.C. in midwinter to give the talk. Nancy sat there, too, listening a few rows away.

The scientist answered questions from people who wondered what they could do to help the planet. One of his suggestions was to live simply, to need less. And to somehow burn less fuel that adds heat-trapping molecules into the atmosphere.

There I thought of the conundrum: gas and oil fuel are still cheap enough that I burn it without much thought. And I live in a place as hard to heat in the winter as buildings are difficult to cool in Phoenix. I often choose to drive to work when I could easily ride or ski the two-mile distance.

After the meeting ended and we had all retreated to our places, I happened to look out the south-facing picture windows of the building, toward the Alaska Range. A sunset washed the pyramid peaks orange at 4 p.m. Another subarctic day faded away.

There, on the packed-snow shoulder of the roadway, a bobbing figure caught my eye. Nancy was jogging home.

13

Longest day

Fueled by the potato chips Jay had handed me, I felt the freedom-thrill of again hiking alone. And I had a goal in mind.

The next day, my friend Andy planned to meet me where the pipeline crosses the Elliott Highway. That was nearly 20 hilly miles away.

As I walked, I felt like a creature created to cover ground on an imperfect-but-effective two-leg setup. Knowing I had a long way to go to meet Andy for supplies and his companionship, I increased both my stride length and my walking cadence. Sweat formed at my temples and under the pack at my shoulders, back and hips, helping to cool my engine. The day seemed perfect, 74 degrees F with fluffy Simpsonsesque clouds blocking the sun every now and then.

That day — summer solstice — was the longest of the

Alaska year. The sun had risen at 2:58 a.m. and would not drop behind the Ray Mountains to the northwest until 22 hours later. The darkest part of that June day would be brighter than the lightest part of an overcast November day.

I had weather agreeable to the human animal, I had the energy, and I had a mission of walking almost 20 miles. That would be a new pipeline-hiking record for me.

Back in 1997, I had walked about 16 miles around Willow Mountain close to Glennallen with Jane, but much of that had been on the smooth pavement and gradual grades of the highway. Now in 2017, I faced a very hilly section of the pipeline over a few decent-size mountains.

The up-climbs on the gravel pipeline pad — some so steep that signs warned "No vehicles beyond this point" — energized me as I powered up them. I floated along on endorphins, reminding myself to save a bit of energy for the last mile. I felt like I was running an ultra.

In the two decades between my pipeline hikes, endurance races had erupted in Alaska and elsewhere. I had signed up for a few ultras (anything longer than a six-hour effort is a definition that seems right to me), and found that I liked them. Why? Part of my satisfaction was that I could run slower in the 43-mile Chad Ogden Ultramarathon on Kodiak Island than I could in the Midnight Sun Run 10K in Fairbanks. I was not running any faster as the years passed, so it was a good fit.

I became addicted to the slow groove: The meditative repetition of one foot in front of the other, the highs that always followed the lows, which followed the highs.

When my friend Ed Plumb invented the White Mountains 100 — a 100-mile run, bike, or ski through a winter recreation area north of Fairbanks — he assigned me bib No. 1 for signing up first.

I chose cross-country classic skis to cover that long loop, finishing 13th in the initial year of the race, during which the temperature dropped to minus 30F in the low spots.

The Ultraski — a 60-mile race along the frozen Tanana River from the town of Nenana to Fairbanks — also became a habit after I skated the initial contest with a guy who carried a gallon jug of water rocking from the top of his backpack. When the promised aid station did not materialize due to a broken snowmachine, my new friend Andy Sterns gave me a few saving hits off his water jug.

I had been running the Equinox Marathon since before the first pipeline hike. I have continued that tradition, ever slower, of running up and over Ester Dome on a September Saturday.

My best financial investment ever was purchasing lifetime Equinox bibs for myself and Kristen when they sold for $125.

The fall celebration is 26.2 miles of slapping hands with friends on the Out-n-Back section, grabbing beef jerky from pop-up homemade aid stations and being handed a triangular knit patch for crossing the finish line.

Three times, in order to keep cashing in on my lifetime bib when Kristen was competing at the front, I rolled Anna over the course in her Chariot; I wanted to go the distance while giving my girl the experience that a friend describes as his Christmas Day.

For the few years organizers offered it, I ran the Equinox Ultra, a 40-mile version of the race held on the same day. It featured another ridge to climb and a few more quality hours on the trail.

Until the Ultra was discontinued — being too hard on scattered volunteers handing out peanut butter and jelly sandwiches for nine hours — I loved the race for its duality: Until mile 20, you ran with the Equinoxers, getting high on that social experience. At mile 20, the Ultra course veered left onto St. Patrick Road. From there, runners were usually alone with their thoughts and their aches for a few hours, until the course featured another Out-n-Back section on the second huge climb — a final opportunity to give high fives to the leaders.

I love endurance sports for the way they make me feel during — floating in a meditative zone — and after: You're done! They make me thankful for my resilient carcass, a gift from both my parents, neither of whom I remember ever being limited by injury.

* * *

My biggest test of pushing those inherited muscles and joints — and my favorite race so far — came when I decided to run the White Mountains 100 course.

It can be hard to wrap your head around moving your legs over 100 miles of country, but I thought it might be possible for a couple reasons: 1) Other people were doing it without dying, and 2) I had finished the Equinox Ultra with the feeling I had something left in the tank.

In reading stories in ultra-running magazines and scanning race schedules printed on the last few pages, it seemed that a 100-miler was becoming what a 26.2-mile marathon had once been — a challenge that was becoming mainstream enough that anyone could run a hundred-miler a month if you had a hard-enough head.

Still, I doubted I could run that far. As I made it into the race off the wait list, I signed up as a runner, knowing that I could, according to race rules, change back to cross-country skier without penalty.

But in the January prior to that March's White Mountains

100, it hardly snowed at all, favoring training runs on local snowmachine trails that got more packed by the day. As that dark month progressed, I made up my mind: I would lay down enough running steps to make me confident my legs could carry me the 100 hilly miles of snowmachine trail through the White Mountains National Recreation Area north of Fairbanks.

One of the best parts of that decision was the training runs. My practice jogs had to be long, really long. I needed to push past my circadian rhythms and keep going when I felt like crawling under a spruce tree. Running a race that long would mean moving through at least one night with no sleep.

My best-executed training run for that White Mountains 100 was maybe my favorite night on my feet ever.

It went like this: First, I gave a PowerPoint presentation at a theater in Fairbanks about a ski trip friends and I had completed from the Seward Peninsula towns of Shishmaref to Nome.

The stage appearance was on a Friday night when I was already tuckered from the work week. The presentation ended in sad/happy tears, as we had lost one of the team members of that trip to a hunting accident. His young widow watched from the audience, looking at pictures of Brian Jackson and listening to his voice narrating video clips. After a few long hugs, I broke down my setup, hopped in my car and drove to Carl's, Jr.

I bought a California burger with large fries and a drink. These I consumed while driving east of Fairbanks into the darkness. There, Kristen and Anna had rented a cabin eight miles from the road in the Chena River State Recreation Area. I parked next to Kristen's car at the trailhead, leaving a plastic sled loaded with my sleeping bag and bivvy sack in the back of the car.

Geared up with a light pack that fit between my shoulder blades, I started jogging up a different snowmachine trail than the one that led to my girls at the cabin. The snow-path, which split from the same trailhead, led me on a 25-mile hilly loop. I chugged along following the bouncing blue light of my headlamp, thinking of Brian Jackson.

Many hours later, I returned to the car after finishing the 25 miles. To the hoots of a great horned owl I imagined was Brian coming back to urge me on, I pulled the sled out of the back of the car. I clipped the sled to a strap around my waist and again started jogging through the night, this time toward the cabin in which Kristen and Anna were sleeping.

In a few hours, I reached thick spruce woods that lined the creek near the cabin. Arriving at 3:15 a.m., I did not want to wake the girls. I instead pulled out my sleeping pad, blew it full of air, and shoved it inside my bivvy sack. I yanked

Kristen's pillowy 20-below down bag from the sled and let it inflate within the bivvy sack.

There beneath the spruce branches I crawled in and felt the delicious reflection of my body heat around my sweaty shirt and ski pants. I knew I would bake my clothes dry by morning. The northern lights danced overhead as my eyes shut. And did a hear a wolf howling?

Runs such as that one gave me confidence I might be able to do the White Mountains 100. But there was the frontier. The race was more than twice as long as any of my training runs.

* * *

Race day in late March started perfect for a runner: Hard-packed trails and a temperature in the high 20s. I shared the first 40 miles or so with my friend Andy, who conveniently skied about as fast as I jogged.

One thing I loved immediately about running versus skiing was that even through running was slower and offered no free ride on the downhills, running was *simple*. No skis to step out of and carry up hills, no pole straps that tended to cut off circulation and make my hands cold. Just two feet, a few pairs of running socks in my tiny pack (which weighed less than 10 pounds even with an insulated quart of water inside) and step after step, no expectation of speed.

That 100-miler felt very long when I reached a checkpoint cabin about 55 miles into the race, 18 hours into my day. My wet feet were on fire, and my ankles and knees ached for a break.

In that warm cabin, 2:30 a.m., my body pleaded with me to crawl onto a plywood bunk. *Just for a moment.* But I resisted, sitting at the cabin's rough plywood table and eating as much meatball soup as I could put into myself while switching into my last pair of dry socks (which hilariously became soaking wet just steps from the cabin as I jogged through water that had overflowed onto the trail).

Sitting down was a concession I avoided when I raced long on skis. Andy had advised me before my first Susitna 100 to Never Sit Down.

During that race, I remember a European skier with sponsor patches sewed all over his ski pants. He was well ahead of me, but I arrived at a checkpoint to see him sunken into a comfy chair. I ate my cheeseburger and drank my Coke standing up, left that warm room before him, and never saw him again.

After sitting for a bit, it was hard to leave that cabin at the far end of the White Mountains loop when I was running 100 miles.

As always happens in long races, a mental up followed that down. For me, it was the sun rising for the second time during my journey, this time firing red the towers of limestone that scratched the air on the far part of the course. My happiness peaked an hour later as a race medic at a tent aid-station handed me a steaming instant coffee in the sunshine.

I drank that Via standing up, but I sat once more on a wooden bench at the final checkpoint cabin 80 miles in. I sipped a cup of Ramen, astounded that I had that many miles behind me. I tried not to think about the 20 miles ahead, commencing with an immediate long, slow climb out of the Beaver Creek valley where cold is born and then, halfway to the finish, the half-mile hill known as the Wickersham Wall.

Dead last among a handful of runners, I was not the final participant. Behind me was my friend (and editor of this book) Eric Troyer. He alternated between pushing and riding a miniature dogsled called a kicksled. Not exactly running or skiing, Eric competed in the skiing division.

Though I thought receiving the Red Lantern Award for finishing last would be cool, I wanted to beat Eric.

I especially wanted this after slogging up the Wickersham Wall. At the top I met Eric's smiling wife Corrine Leistikow. She is also my encouraging doctor, who, when I asked her if it was really a good idea for anyone to try and run 100 miles (hoping she would talk me out of it), responded "Of course!"

Corrine had news for me at the top of the wall, about 8 miles from the finish.

"Eric's not that far back. He's catching up to you!"

I peered back into the fading evening light and could not see a dark spot coming up the wall, but her words had fired my adrenal system.

Though I knew Eric was pushing his kicksled slowly up the Wickersham Wall, on the downhills to the finish (there are several) his rig would be twice as fast as me. This was a race!

As the darkness of the second night ascended, I entered a physical zone I had never reached before and may never again: With 90 miles in my legs, I pushed for all the speed I could muster.

I lengthened my stride on downhills, imagining stars and crescents shooting from my knees and ankles. I put my head down on uphills and sucked in cold air like a locomotive burning coal, firing my played-out quads. To my amazement, they responded.

I felt like I was running fast, though I probably managed only about three miles per hour. Still, I was performing the running motion with almost a Franklin on my pins!

Then, with no Eric in sight, I reached the top of what I knew to be the last hill. I saw a sign attached to a wooden stake stabbed in the snow:

1 MILE TO FINISH ! ! ! (MILE 100)

I turned and looked one last time behind me. I could not see Eric's headlamp, just the dim outline of boreal forest with a bluish cut line of trail.

One of my goals was not to turn on my headlight for a second night's travel. I left it in my pocket even though it was then nearing 10 p.m. (my second 10 p.m. of the race) and more trail light would have been a good idea.

In those last hundred steps, I experienced a euphoria. Thanks Tony (dad)! Thanks May (mom)! Thanks Kristen and Anna (for putting up with my hours of training)! Thanks to Whatever Entity up there looks over people who do dumb-but-fun things!

A smile on my face, I jogged down to the finish line at the trailhead parking lot, suddenly self-conscious in my stinky and snotty state about meeting the race volunteers I know. I crossed a line of spray paint on the snow to the sound of mittened applause and admired some neon graffiti someone had sprayed on the parking lot.

Then the artwork disappeared, and I noticed Andy

standing there. He had dropped out after about 40 miles. After a snowmachine ride out, he went home to nap and had returned to drive me home, so I didn't kill myself by falling asleep on the highway.

Andy handed me a cheeseburger that Kristen had bought for me. She, on skis, had finished the race almost a day earlier and was home sleeping with Anna. The burger, cold and congealed, was the best thing I've ever eaten.

Just 16 minutes later, as I stood there in a happy stupor, Eric came sliding in on his kicksled, stopping before he hit gravel in the parking lot. His euphoric smile matched mine. The White Mountains 100 was over.

* * *

Let's see. The pipeline is 800 miles long. I covered 100 miles on foot in 38 hours during that race ("You were on your feet for a work-week," a woman told me). So why couldn't I bust out the pipeline in a couple weeks?

For one thing, it took a few weeks to recover from that century run. Ski coach and friend John Estle has told his charges that to recover from an ultra your body needs one day of rest for every mile you ran. Plus, I don't think I would remember much of the route if I covered in in a few weeks rather than a full summer.

Unlike the White Mountains 100, the pipeline hike was much more about the journey than the finish line.

And, man, was I enjoying summer solstice. Cora and I topped a hill with a view of the trough of the Tolovana River valley. There lay another of Alaska's great green swamps, this one hiding the self-healing Minto Flats seismic zone, which shakes us with regularity in nearby Fairbanks.

Walking down the hill, Cora and I eventually gained the flats of the Tolovana River valley. Amid the black spruce and wet moss, the pipeline's 800-mile gravel road was essential to foot travel; without it, we would have been slogging through black, knee-deep boreal muck.

Soon, I saw what I knew was coming: an orange sign with a black number 400 attached by a bracket to a pipe support. On night 47 of the trip, we stood equidistant from both Valdez and Prudhoe Bay. Four-hundred miles down, 400 to go.

We were halfway, but most of our peopled journey was behind us, in the towns of Valdez, Copper Center, Glennallen, Delta Junction, North Pole and Fairbanks. Ahead were Coldfoot and Wiseman and lots of quiet country with no cellular waves. Yay.

At 10 p.m. on summer solstice, golden light torching the willows and spruce, Cora and I reached a wooden bridge over the Tolovana River. I pitched our tent beneath the spruce trees just off the bridge, and then walked back to the wooden deck for a late dinner. We liked the bridges because they fooled the mosquitoes a bit and provided a flat platform for the stove.

Near midnight while I hung our food for the night, my satellite tracker jabbed me in my pocket. I remembered I hadn't turned it off. I pulled it out and looked at our distance — 19.4 miles. Our longest pipeline day. On the longest day of the year.

14

Isom Creek

"BLOCK POINT: WINTER ACCESS ONLY."

The orange-and-black sign fronting the rock-and-soil berm gave warning of the deep, dark valley behind it. Our easy walking was about to get hard.

Cora and I had passed several identical markers during the last few hundred miles. Block points are berms of soil and rock along the pipeline's 800-mile route meant to prevent vehicles from going any farther.

Beyond the berm, pipeliners built no gravel road alongside the pipe. Sometimes the permafrost beneath the surface was too close to thawing; engineers opted not to blade the insulating ground cover. Sometimes a hill was too steep even for bulldozers to build a road.

The descent before me likely had permafrost beneath. The engineers had good reasons for not extending the pad here, but I was not thrilled with the spongy muskeg walking to come. The annoying, unsure footing would slow our progress through the shaded valley below, which was choked with black spruce, a sluggish creek and billions of whining abdomens connected to syringes.

It was 7 p.m., about time for the bugs to get more active as the day cooled. Where was Andy?

Andy Sterns had joined me a few days earlier, delivered at 9:20 a.m. by Joe Wagner, who had been driving a van full of tourists north to the Arctic Circle highway pull-off.

Where Andy had met me — at the pipeline's crossing of the Elliott Highway where that road continues to its conclusion at Manley Hot Springs — was exactly where he had joined me on the first hike 20 years earlier.

Many elements of our lives had changed since 1997, but our "Kevlar friendship" — Andy's words — remained bulletproof.

I wrote a chapter on Andy in *Walking my Dog, Jane*. People said it was their favorite. In that book, I described the cross-country skiing accident Andy suffered that so affected his life.

Andy, a member of a Vermont collegiate team, was skiing with a friend during a cool-down loop after a race when he skidded on an icy corner. He carried his speed into a tree, hitting it head-first. He suffered a compression fracture of his fifth cervical vertebra. In other words, he broke his neck.

Some doctors said he would never walk again, but Andy rehabbed the hell out of that injury, pushing himself hard, first out of the wheelchair and then off his crutches.

He did not return to 100 percent. He could not pick up his right foot, and still needs to hoist his leg with his hand to get into his car. But Andy has learned to compensate.

Now when he walks or runs, he slices the air with his bent arms parallel to the ground, as if he is trying to work out a kink in his back. The loss of mobility has robbed him of ground speed — he had been a fast runner, biker and skier — but it hasn't slowed him down. He has accomplished things that 99 percent of the human population will not.

Mush the Iditarod, twice. Skate ski 674 miles from Nenana to Nome. Ski 160 miles in self-supported winter wilderness races. Ride mountain bikes on snow with two Canadian uber-athletes, from Dawson City in the Yukon Territory to Nome — more than 1,000 miles, a hundred of them at 40 below zero.

The list goes on, but Andy will never recite it. A Fairbanks newspaper writer once nailed him with this lead to a profile: "Andy Sterns has no rearview mirror."

Citing Andy's quirks — such as his inability to ration food because he lives so utterly in the moment — one of the Canadians who biked to Nome with Andy said, "He is one in six billion."

A mutual friend summed Andy up with my favorite: "Nobody does more with what he's got."

I felt guilty that Andy was not in sight as I climbed over the dirt pile and began the tundra descent into the Isom Creek valley. I justified leaving him behind by reminding myself how difficult it was for me to walk at half-speed after my legs had moved a constant rate for the last 50 days. Once, while we walked together, I checked my GPS: 1.8 miles per hour. I am pretty slow, but alone with Cora I averaged more than 3 mph.

I winced as I pressed down on the bouncy vegetation that led down to Isom Creek. It was slow going. Looking ahead, I could see where the pipeline finally climbed out of the spruce valley. The hilltop on the far side of the creek — where we would camp — lay two miles away as the raven flew, and we didn't have wings. Getting there would take Andy forever.

The lowlands, thick with wind-blocking spruce trees, were Mosquito Paradise during June. Now June 25th, the clouds of skeeters swarmed thick, each insect buzzing madly toward blood-protein sources during its few weeks to shine.

The camping-on-hilltops strategy worked often enough, but it came with a disadvantage. By mid-summer, spring snowmelt has drained from much of the high country. There is usually no free water on top. That's why I carried collapsible jugs, which Andy had purchased for me a few weeks before.

Having scratched my way downhill, I dropped my pack next to Isom Creek, a liquid groove curving through brown mud. Cora lapped the cool water with her ladled tongue.

Before digging for the water jugs stored deep in my pack, I pulled a tiny pump bottle of DEET mosquito repellent from my left thigh pocket. I sprayed the back of my left hand, then wiped the back of my right hand on it (I tried to avoid getting the bitter liquid on my fingertips or palms, where I might eat it later). I rubbed it into the skin of my hands like sunscreen. Then I spritzed my baseball cap, neck, arms and ears. I pulled out the zip-on lower sections of my pants even though I knew I would sweat while ascending the steep hill out of the valley. Shorts felt nice, but it was time to don the armor.

Cora, blasting through the bushes, shed her own mosquitoes by brushing them off with branches, so I didn't dope her up.

I found both jugs in my pack. I pulled them out, then

dunked each in the black water of Isom Creek, small enough there to step across. Each container held 1.5 gallons. Three gallons, I hoped, would be enough to hydrate our dinners as well as morning oatmeal and coffee, with enough left over to keep Cora alive. Then I bent down, grabbed the handles of the water containers, lifted the new 25-pound load, and started up the mushy incline.

Soon, the pipeline pad reappeared, making me happy. The gravel road was barely used, grass shoots popping up from the tire ruts, but it was much easier walking than the moss and Labrador tea plants.

Cora and I started hoofing the mile through the flats of Isom Creek toward the hill that would be our camping spot. I could not manage the 7 mph needed to outpace the mosquitoes, but it felt good to move.

With my hands occupied, I was thankful to be bathed in repellant. Most of the mosquitoes seemed to bounce off me, confused by the DEET's carbon dioxide-masking molecules.

As I shuffled up the mile-long climb to a plateau I had seen on my paper map, I noticed familiar indentations in the mud on the side of the trail. A bowl that resembled a human palmprint, with four chisel marks at the leading edge. Grizzly bear. Fresh.

Cora stuck her nose in the print and sprouted a Mohawk,

black hair standing on head and tail. I started singing a King's X song to be a little more obvious.

Andy was back there somewhere, beginning his jerky descent into the valley. I reminded myself that he was lucky to be there. To be anywhere.

* * *

"He's fucked up," Ian McRae, still trying to make sense of the accident, said over the phone from Nome. "It's a game-changer."

Four years earlier, in April 2013, I had called Ian after hearing of Andy had broken a leg in a climbing fall north of Nome. The initial report I heard was wrong. Andy had broken *both* legs.

"Both set of tibs-fibs," Ian said. "Snapped like twigs.

"He's at Providence (Hospital) in Anchorage. I'm not sure how he's doing but I think I'll visit him in a few days.

"I've just been hunkering down and — you'll understand this — drafting a blog post on the accident. It's the only way I can process."

Ian, a mountain climber my age with Popeye forearms, teaches fourth graders in Nome. I imagine his students like him because he is funny and seems kid-like himself. A California boy, he had migrated to Alaska and married the daughter of an iconic outdoor-gear shop owner in Fairbanks. They moved to Nome more than 20 years ago.

One of Andy's many climbing partners, Ian had invited Andy on a spring attempt on north face of Mount Osborn in the Kigluaik Mountains, Ian's semi-private playground on the Seward Peninsula.

At 4,714 feet, Mount Osborn would be a respectable Adirondack peak, but is not a goal of many western mountaineers. Still, Ian saw a worthy challenge in a sheer 3,000-foot wall that ran up its north face. He had scouted it over the years on snowmachine trips from Nome, about 35 miles away. As far as he could tell no one had every climbed the peak via that route. He and Andy had tried before a few springs earlier, but according to Ian had retreated, due to "severe spindrift."

The climb is no walk up. It requires ropes, helmets, crampons and most everything you would need to climb Denali. But Ian and Andy were prepped for the challenge. They had the skills and the equipment. All they needed was a little luck.

* * *

I had not seen Andy in a bit when he accepted Ian's invitation. A short time earlier, Andy and his longtime girlfriend Lisa had broken up. In the aftermath, he threw himself into

SUMMER OF GRAVEL AND STEEL

one climbing endeavor after another: Rocks within driving distance of his parents' home in Massachusetts, ice in Vail with a friend from Denver, and Grapefruit Rocks and everything else close to Fairbanks.

In those exploits we differed. I have almost no mountaineering experience, except for a trip up Denali once to help fix a weather station at 19,000 feet. There, with a Japanese team, I puked up some wasabi crackers and descended to high camp at 17,200 feet. I never returned to try for the summit that was just one mile of non-technical hiking away.

It was not my country. Most of the time, I wondered what I was doing up there standing on snow while looking down at the inviting blue-green Kuskokwim Flats where the temperature hovered in the 80s. The altitude made me feel fuzzy all the time. Despite living outside for three weeks, the experience was surprisingly claustrophobic.

Andy — with ant-like upper-body strength that allows him to compensate for legs that do not work as they once did — loves the high country. During our shared adventures, he has never been scared in the high, windy places that terrify me.

Andy has other partners for those exposed endeavors. He had accumulated a varied team of climbing chums who called him "Everyday Andy" for his extreme availability. Andy has never worked a regular job since I've known him. I've never asked him how he pulls it off, but his low-overhead lifestyle helps.

Andy and I had drifted to different bays during that period of our lives, after a previous decade of adventures that always included a spring ski trip, like from the villages of Tanana to Allakaket.

But we continued to check in with one another every month or so. After he said yes to Ian, Andy called me while waiting for his flight to Nome in the Anchorage airport.

He told me about he and Ian's previous attempt on Mount Osborn, and how their weather window seemed better this time. They would ride double on Ian's snowmachine from Nome all the way to the base of the mountain, where they would camp. There was very little hiking to get to the place they would start climbing. Perfect.

As Andy hung up, I thought of the heavy hitters Andy had recently climbed with. To put up with his slow approaches to climbs, I figured they must appreciate, like me, his cheeriness, decisiveness and how much he loves being out.

* * *

Sweating, I reached the top of the hill out of Isom Creek and I set down my pack, feeling a chill at my lower back. I wondered how long Andy would be. Because Andy can do so much on legs held together with titanium pins, it's easy to forget the accident on Mount Osborn ever happened.

But there I took a moment to think about it, and allowed myself to appreciate how he was still on the move after two major accidents, both of which could have left him living an indoor, immobile life.

I found an acre cleared on the hilltop and pitched my tent; I tossed my baseball hat on a similar site right next to mine to mark a place where Andy could set up his tent. That decision can take remarkably long sometimes, especially when you are tired.

Later, when it was nearing 11 p.m., I sat waiting for Andy. I could see a long way, as it was nowhere near dark only a few days past summer solstice at this latitude. Clouds took on the pink hue of played-out sockeye salmon.

Where the frig is he? Did that bear eat him?

I wasn't really worried about the bear. Andy knows how to take care of himself in the bush. But as the minutes ticked on, doubt started to creep in.

Just then, I saw a blob of yellow firing up the pipeline road. Andy was ratcheting up the hill, wearing his yellow raincoat because mosquitoes couldn't bite through it, head bobbing in time with his trekking poles.

When he arrived at the campsite, many minutes after I first saw him, Andy let loose his pack; it landed with a thud. I got some water boiling for his dinner. Cora and I had eaten an hour earlier.

Nearing midnight, as Andy drank the sauce of his pasta primavera and I sipped a cup of tea, a great-horned owl lit in the top of a spruce tree right in front of us. The spruce swayed with its weight. We both saw it blink.

"Cool," Andy said.

* * *

From Ian McRae's blog:

I had just finished equalizing the two screws when the rock came. They always seem like such a presence, the Dark Knight suddenly flapping into your airspace, they are sensed with the sixth, not the vision or the hearing.

I took hold of the power-point 'biner and cowered under my helmet. The Death Eater passed close over me. A huge snow avalanche followed that suffocated and threatened to sweep me.

Andy was swept, which I was able to infer when a tug of force came on my harness, which in my new small-diameter ropes freshly ordered from A.M.H. (Alaska Mountaineering and Hiking) absorbed magnificently well. I was waiting for something hard to intersect me out of the snowy slipstream, with that weird stupid f-ing surety one gets that one is certain to survive this dreadful thing that has come.

The avalanche stopped. The screws had held. Andy was screaming bloody murder a full ropelength below. The blue rope was core shot for twenty feet, but the kern was intact. The orange rope was strangely buried in the snow. There was tension on the rope and on the screws, and I could not immediately meet Andy's frenzied request for "SLACK, PLEASE!, ON BLUE!!"

* * *

High above Isom Creek, Andy pitched his tent without a headlamp during the darkest part of the evening, which was still light enough for him to see the openings in pole sleeves and the tiny zipper tabs. There, at midnight, the hilltop was cool, hospitable to mosquitoes.

I lay inside my tent with Cora pressed against my bag. Through the protective mesh, I watched as my friend unzipped his tent door and chucked in his stuff. Nothing made it on the first try. He picked up his sleeping bag and slung it toward the tent opening; it slipped out of his hand and thudded before the entrance. He picked it up again, dropped it, and finally tossed it inside.

His sleeping pad and pack went inside with similar delays. *I could do that so much faster*, I thought. Yes, and so could any of Andy's other friends who stick to the side of a mountain with him.

We are patient with Andy (Ian referred to me as "Saint Ned" in blog posts) because the only other choice is to complain about his slowness, and what good does that do?

Many times, after he has hit the ground next to me, I have asked Andy if he is OK. He always says yes, even when his nose is bleeding. So, I have stopped asking. On the rare occasions where he has not been OK — such as when he hit his head on ice while skiing and suffered a concussion — we assess things and turn around. But all Andy's friends know he can take a lickin' and keep on tickin'.

I have resolved not to help Andy unless he asks me. And I have learned to savor the extra time on winter trips when Andy prepares himself to move each morning. I brew an extra Via and look at maps.

Finally, above Isom Creek, Andy lurched into his tent. As he zipped the door shut, I heard him slapping some of the 46 mosquitoes that entered with him.

***

Both legs were broken. There simply wasn't any place on the scale for the pain Andy was in . . .

We don't know what happened to bust Andy's legs. Either the rock actually banged him, or more likely, Andy's legs got tangled up in a loop of slack during the 60-foot fall; when the ropes came tight, ouch.

He reported to me that he didn't feel pain until the fall-catch.

* * *

His legs. Pencil thin and willing but largely unresponsive to the signals coming from his brain, they shuffled forward, leaving parallel crescent scrapes on the fine gravel of the pipeline pad. Sometimes I walked ahead to get audio relief from the crunch of pebbles being pushed, pushed, pushed.

Twenty years earlier, only his right foot dragged. The accident on Mount Osborn had taken other abilities.

In the late 1990s — between accidents — skating was Andy's preferred of skiing. He and I shared that technique while blocking wind for one together during the first Ultraski race, a 50-miler along the frozen Tanana River from Fairbanks to Nenana.

While following him, I noted a milk jug full of water hanging from his backpack. Later that day, he let me sip from that container when the promised aid station was not there at mile 25 (because the snowmachine broke). That's how our adventures together started.

Many of them involved skate skiing, the longest being a 26-day trip down the frozen Tanana and Yukon rivers and then overland to the Bering Sea coast and Nome.

Due to his college-skiing accident, he could lift his right foot only a few millimeters. That forced him to pole solely to one side (skiers call it V1 to the right). But Andy could do it for hours and hours, like a wolf trotting the hills.

Mount Osborn removed skating from Andy's toolbox. After that injury, on skis he is now only able to double-pole, a technique in which your upper body and trunk does most of the work. Double-poling can be fast on the flats, but takes more energy and doesn't allow for much recovery. No one can double-pole like Andy, but on hilly courses he opts for sticky wax tape that helps him advance upward but robs him of glide.

Post Mount Osborn, Andy struggled more with walking, too. But he *loved* the simple pipeline life of ticking along with hiking poles and tiring himself out while thinking of pasta primavera. He was smiling every time I looked over at him.

In the 20 years since we had last covered those hilly miles north of Fairbanks, Andy had remained single and unattached, still owned no furniture other than a few gifted wooden chairs, and slept inside his duplex on a foam-rubber pad. His eclectic eating preferences had shifted from cabbage and Power Bars to gas-station apples and Kind Bars, but a constant has been a love for mustard. Kristen and I once gave him a two-gallon commercial pump barrel of French's for his birthday.

* * *

Got to the bottom of the route after hours of tedious lowering . . . Merciful gods, no more gravels or snow came down that wretched slot that day, my eyes glancing up there like the hunted, Andy's eyes glazing with shock, shock, shockamundo. We would have climbed it.

A long and horrible drag to basecamp began, 800 feet down to the erratic. Andy got going backward, me dragging him through the snow by his daisy, and using the hood of my Wild Things Belay Jacket under his head as a keel . . .

His Joe Simpson legs tippled along on their heels obediently behind him like merry little broken sleds. He was horrendously hypothermic and shivering violently, but able to inform me of this fact, which he did over and over again, which I told him was a good sign.

* * *

Andy studied geography at Vermont's Middlebury College, where he ski raced. He knows the Alaska map even better than the Vermont map, because he's travelled all over this big peninsula. His trail savvy had us sleeping indoors most nights on a 350-mile ski from Knik to McGrath, because he knew where Slim's Roadhouse and other places were. When we squeaked into those cabins brushing snow off our pants, the proprietors knew Andy's name.

In our first Alaska Mountain Wilderness Classic Ski Race from Nabesna to McCarthy, every molecule in me wanted to follow a fellow racer's tracks up Flood Creek. That chap had skied in the direction we wanted to go, I thought.

"I don't think it's right," Andy said as he squinted at the map (this was a few years before GPS hit the scene). "We've got to keep going straight."

"But Brian went this way, and it's headed toward McCarthy," I shot back. "How do you know it's not right?"

He pointed on the map to where he thought we were, following a skirt of gravel around a receding glacier; the cut of Flood Creek was indeed pointed toward McCarthy, but led to a mountaintop from where travel would be impossible.

I was skeptical but unsure. Andy was sure that Brian had made a wrong turn. I gave up, and agreed with him to keep moving.

Brian had to be rescued a week later after retreating from a mountaintop. His broken tent might be up there still.

Andy and I had shared enough of those moments that eventually I left the navigation to him, which allowed me to groove on the landscape, take photos, daydream. One of our greatest shared pleasures was passing the paper map between us at dinnertime, talking of the next day and what we might see.

* * *

I had forgotten the SPOT device outside the tent, the two green lights like twin owls blinking away synchronously on 911 . . . futilely, we had come to believe. Yes, people of the New Age, in the quick of the thing Andy and I had lost faith in the BUTTON and reverted back to the wilderness of our youth, when no one had a GPS . . .

But little did I know, little SPOT guy had successfully, and rather randomly I might add, slipped the 911 signal around the bulk of Osborn . . .

Andy heard it first. "Helicopter!" A Bering Air red Robinson, with Cory at the controls. It soon disgorged Wes from Norton Sound who wasted no time crawling into the stinking mayhem of our tent to administer to Andy. Broken legs confirmed. In went the I.V. On went the splints . . .

I don't know anything at the time of this writing about Andy other than they medevacked him to Anchorage.

"It must be nice not to be in constant pain," Andy growled.

I looked up, stunned, from the book I had been reading aloud. Andy, in a coma for a week, had strung together a real sentence.

After the accident, Andy had been flown from Nome to Anchorage. He had undergone surgeries to repair his legs. Doctors said the breakage of four large bones — both tibias and fibulas in his lower legs — and the 30-hour delay until he received medical attention had caused marrow leakage and "a fat-clot shower to his brain."

Fat from Andy's freed bone marrow had seeped into his bloodstream. These fat-clots, called emboli, can obstruct blood flow in the brain and lungs. The doctors believed the fat emboli were causing swelling in his brain, and that it would fade as they dissolved. At least, that's the theory. Medical experts don't know how it all happens, which makes it even scarier.

"We just need to let his body rest," one doctor said. "He should have a couple unresponsive days like this and then slowly turn around."

Andy's unshakable mom Judi arrived from Dartmouth, Mass., for the long vigil at Providence Hospital in Anchorage. She had company for dinner most every night as Andy's friends, many who lived in Anchorage, came to visit the seemingly sleeping Andy in his hospital room. Andy's dad Bob, another calming presence, soon followed Judi.

I traveled down for a few week-long shifts. During one of those, Andy started to speak. When in Anchorage, I stayed for free at a home owned by emergency room doctor Andy Elsberg and his wife Shannon Brockman. Andy Sterns and I had met Andy and Shannon in the windy village of Shaktoolik on the Bering Sea Coast. We were skiing the Serum Run route from Nenana to Nome at the same time Andy and Shannon were mushing two teams of dogs on their honeymoon trip.

To have a doctor like Andy Elsberg at our side was a great comfort in a time when we didn't know if Andy would ever speak again. Maybe that rugged little body had finally been whacked too hard.

But it only took a week before Andy began stringing sentences together. The real turning point was when Andy asked for mustard to smear over his hospital meal. He gobbled up mashed potatoes with three plastic packets squeezed upon them. He was also soon requesting the strong coffee from the machine in the hospital lobby, another excellent sign.

While Andy eventually healed enough to tackle outdoor adventures again, he isn't up for the multi-day winter adventures. The loss is more his, but I feel it too.

I have mourned that Andy no longer cooks up springtime trips for the two of us. No more grinding along cold snow squinting to see the chisel marks of dog toenails along the old mail trail to Allakaket. Or sleeping at the cabin on Big Lake with the six-inch-square window above the door and the handbook on Alaska trapping left on a bunk. No more calling the Galena school to see if we can crash on the floor after we clank up the metal grate steps on some random night in March. Or hoping there is dry wood to be cut near Old Woman Cabin before the push to Unalakleet.

Age was pulling those trips away from us anyway, but Mt. Osborn accelerated the process. There are still adventures to be had, though. Like hiking a gravel road together.

* * *

After walking through a corrugated-metal tunnel that ushered the pipeline under the Dalton Highway, Andy and I emerged back into daylight at the Yukon River Bridge.

Before us was a drawbridge-like structure one-quarter mile long, decked in Douglas fir boards from old growth Oregon trees.

Andy stabbed his poles into the wood near the downstream side of the bridge, the one with the view of the blue-green Ray Mountains in the distance. Twenty years earlier, with Jane, I had paused there to take one shot with my point-and-shoot camera. When I had it developed a few months later, I saw to my delight that I had captured a bolt of lightning sizzling to a mountaintop.

There was no threat of lightning when Andy and I crossed. It was a hot day in the middle of the warmest part of Alaska, with steaming bogs below and blue sky above.

On the deck of the bridge, I secured Cora to her leash and that to my waist buckle, hoping not to hear truck engines. Drivers would be moving fast up the bridge, pulling lighter loads on their return from the north.

The Yukon's coffee-with-cream surface flowed 100 feet beneath us. Aprons of brown gravel on each shore suggested the river level was dropping on the 80-degree day. It still carried an unimaginable amount of water past as it snaked left around a bend downstream. Where does it all come from?

Reaching the far side, we peeled off the bridge to the left, on the downstream side. We both dropped our packs next to a pair of aluminum river boats, one with a Honda 4-stroke and another with its outboard cowling removed, its owner right then off retrieving a part. The propeller of that motor was tilted out of the water. It was painted black except for where the silver leading edges had the paint chipped off by unseen rocks.

Cora jumped in Alaska's longest and widest waterway,

happy to fetch a willow branch I had tossed. I soon joined her, stripping down to my underwear and stepping into the eddy current that shoved us gently upriver. I walked farther in and opened my eyes underwater, imagining a sandstorm in the Sahara.

When I lifted my head above the water, I was surprised to see Andy — who wears two wool ski caps on summer nights — kicking off his pants. Was he joining me in the cool water?

In his boxers, Andy shuffled into the Yukon and knelt until he went under. He popped up, rivulets draining from his nose and chin, baptized by the Yukon.

He stood there a second, half his body invisible in the river, and smiled. Andy was pure muscle, a cabled instrument of movement. His ripped abs, pecs, shoulders and biceps were ghost white, his forearms a chocolate brown. The Alaska farmer's tan.

Andy in the Yukon River

He would soon leave me there at the Yukon crossing. Another friend would join the walk for a bit.

As he cleared the water from his eyes with the back of his hands, he turned to face the way we had come. For a second, I thought he might be reflecting on the 100 miles we just covered. A look in the rearview mirror?

Then he turned around and looked northward to the green hills still to come.

"You and John have some nice ground to cover," he said. "Thanks for letting me come along."

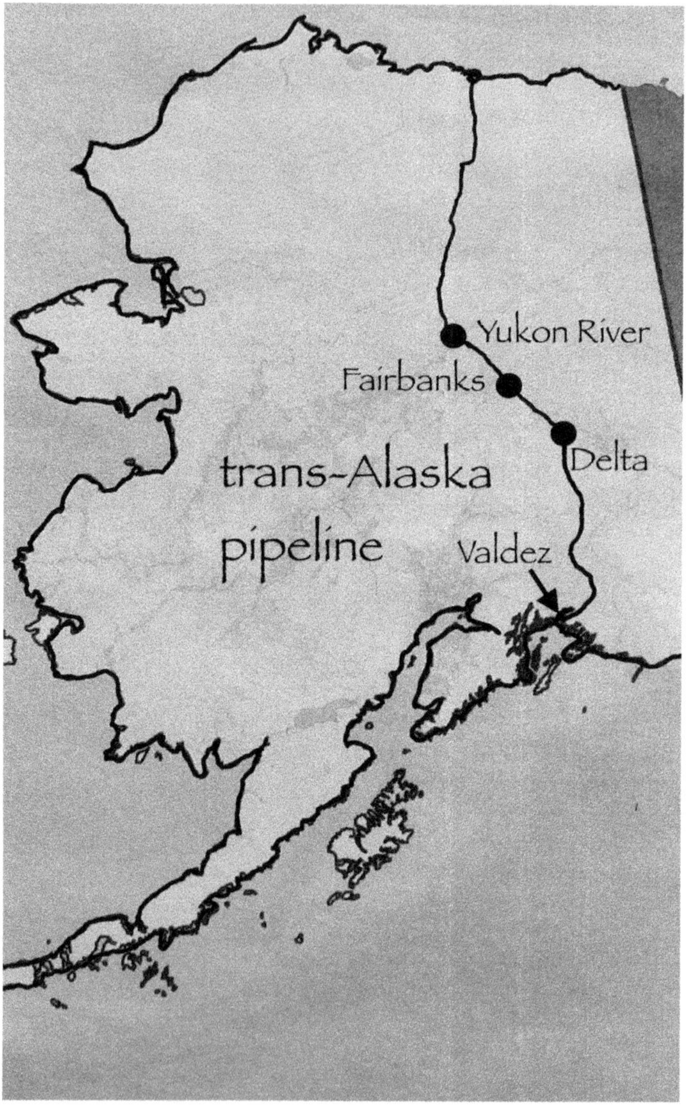

trans-Alaska pipeline

- Yukon River
- Fairbanks
- Delta
- Valdez

15

Yukon River

The Hudson River of my youth flowed less than a mile away, down the hill from our Main Street home in Hudson Falls, New York.

I would walk down the railroad tracks and over the Fenimore Bridge, where the river plunges over a broad shelf of shale. Crossing the bridge, I smelled the musty froth of mist, pounded into the air by that freefalling water.

Looking back on the Hudson with Alaskan eyes, the river crashing through my hometown is like nothing I have seen up here. One-hundred fifty miles downstream from my hometown, the Hudson is more than three miles wide and 200 feet deep as it makes its way toward New York City. I'm impressed with its massiveness whenever I'm escaping my sister's house in Brooklyn over the George Washington Bridge. No wonder

Henry Hudson steered the *Half Moon* up that slot in the 1600s, when he was searching for the Northwest Passage.

Henry gave up in Troy, New York, where I lived before driving to Alaska in 1986. There, more than 100 miles upriver from its mouth between Jersey City and Manhattan, ocean tides still make the river rise and fall.

In my upstate patch of pine woods near the Hudson River, I shot a few grouse with my father's 20-gauge shotgun with the choke you could adjust with a twist. I sometimes fished the river eddies, but my mother told me not to bring any of the fish home. She had been reading in the *Post Star* newspaper about how officials for the Environmental Protection Agency had required General Electric to remove pollutants from the river, starting at our town all the way down to the big city.

One General Electric manufacturing plant was in Hudson Falls and the other in Fort Edward, just south of us. Parents of my friends worked at the factories. Black-and-white photos show them standing over large vats, holding things elbow-deep beneath the surface of a liquid with a sheen on top.

My friends' parents made electrical capacitors that ended up high on power poles and inside radios, TVs, and other things we liked and used. Within these components large and small was oil. That oil contained chemicals known as PCBs, which scientists later found to be carcinogenic. For a few decades, managers at those GE plants on the Hudson released more than 1 million pounds of PCBs into the river. Some of the workers later sued, claiming that working with the PCBs made them sick.

Now, many years later, those PCBs are held in the sediment of the river bottom near my boyhood home. That's where environmental stewards figured they would do the least harm, after years of dredging out the contaminated mud, plopping it in a truck and storing it on land.

The Hudson River tugged at little Ned who needed to be outside. But I couldn't drink the river, or eat things I carried home from near the shore. Knowing that didn't keep me away, but it no doubt affected me.

* * *

Andy caught a long ride to Fairbanks in the open bed of a pickup truck. He hitched home with a woman named Cheryl from Stevens Village who had just been down at fish camp, scooping up chum and Chinook salmon at "the Rapids" downriver of the bridge. It's a good bet that the fish in Cheryl's coolers did not have measurable traces of PCB in their flesh, though they probably had ingested microscopic bits of plastic that are now ubiquitous on Earth. A researcher in Juneau recently found microplastics in every single rain sample her students collected in the mountains and waterfront around Alaska's capital city.

Awaiting my next hiking partner, I set up my tent at a little designated site just downstream of the bridge. The walk-in gravel pad featured a fire grate and a picnic table with a view of the Yukon. For water to boil for dinner, I dipped my pot in the Yukon.

I fed Cora dinner and unrolled her pad, a RidgeRest cut in half. She curled up under the picnic table at my feet. I sat down and stared at the river.

The brown Yukon flowed by, tabletop flat and a half-mile wide at this midway point on its 2,000-mile journey. Whirlpools dimpled its otherwise smooth face.

The river cuts the main body of Alaska in half. To the north, the last stubble of boreal forest on the continent. To the south, the few Alaska cities, a couple big mountain ranges, and all the volcanoes.

Sitting there, I flashed back 25 years to a different point on the Yukon. A Japanese kayaker sat at a picnic table in a campground just across the river from Dawson City in Canada. He sat there like a statue, without a turn of his head or a mosquito slap. I would walk to the ferry, ride its arc over in the current until it crunched the far gravel bank in Dawson, lose some loonies at Diamond Tooth Gertie's and stumble back to my tent in the dusky midnight light. The Japanese guy was still there gazing at the water, Zen waves wafting off him.

The Spell is powerful. At my campsite near the bridge, I felt it, even if trucks barreling down the bridge made me turn my head that way.

Watching all that water on its journey to villages downstream and then the Bering Sea can put you in a meditative state. That belt of smooth water occupies the mind and — like a calming blanket settling over the mind-monkey — the soul.

The monkey interrupts the meditation: One-third of all the flowing water in Alaska finds its way to the Yukon. It's a number impossible to comprehend, but easy to believe while sitting at the picnic table. Even here, there is so much more water to yet to join the river: Downstream is the Tanana (itself 200 miles longer than the Hudson), the Koyukuk (even lengthier than the Tanana), the Melozitina and the Nowitna. And so many more.

Where does all that water come from?

The river's headwaters are high-mountain lakes of British Columbia. They release pulses of snowmelt, rainfall, and ice melt when spring temperatures rise above freezing. The Yukon collects the gravity feed of water from most of middle Alaska and the Yukon Territory, until much of winter's snows have melted by late June.

Melted glacier ice swells the river to its banks in early June, and much of that seeps into the ground to become groundwater that continues to feed the river. The Yukon drops far from the spruce in autumn, when the mountain streams harden again.

The sole bridge in Alaska is the one right there to my left, slanting downhill from the south to the north side of the river. Built to support an industry that enables the state to pay its bills.

In this way, this bridge has attracted workers and has kept them here with Alaska Permanent Fund Dividend checks. Not far from here, a worker for Alyeska Pipeline Service company told me that at the peak of the pipeline's oil flow, he once stood above the buried pipeline. There, he calculated his annual salary in oil passing beneath his feet in seven seconds.

A proposed dam was never built just downstream at Rampart Canyon. One of Alaska's Magnificent Swamps — Yukon Flats — was not flooded to provide electricity for Alaskans. Millions of ducks and geese and songbirds still touch down in the flats every spring to have their babies. They need land to build their nests. If the dam had been built, there would be fewer new birds, my electricity would be generated by water rather than coal, and this picnic table would be 20 feet underwater.

Remembered times on the Yukon:

- First ride in a heavy aluminum riverboat as a 30-year old park ranger working out of Eagle. Squinting at the wind and the silt particles hitting my eyeballs as they spiraled off the floor of the boat. Stopping at the renovated cabins of the miners and river rats who built elaborate dog runs out of peeled spruce posts they had pounded into the flour soil. They were going to stay forever.

- Meeting all the Eagle characters within *Coming into the Country*, the John McPhee book my father mailed to me when he learned I would be stationed in Alaska after finishing Air Force technical school. John Borg's wire brush of a moustache and Dick Cook's need to tell stories of the Bush after he splashed upriver into town by motorized canoe. Spot on.

- Years later — 160 miles and one week long from Eagle to Circle — the sandy, bugless island we dragged the red canoe onto during Anna's first river trip. At three years old, cheeks painted with sunscreen after losing her hat to the wind, playing in the river with a plastic pail and shovel. Kristen sunbathing on a towel nearby, squealing when Poops shook a spray of Yukon onto her. The black dog splashing in again and again, fetching a curved stick of birch, until, head swiveling to all compass points, she finally lost sight of it in the current. The birch stick continued on downriver, like the casket of the Canadian who left a note to please push him out in the current should he be stuck in an eddy. He wanted to reach the sea. One Eagle friend telling the tale that he — with his back to a log using both feet as if doing a leg press on an exercise machine — shoved the casket back into the current.

- One May years past, traveling to this very spot near the Dalton Highway bridge to camp with Anna and Poops while Kristen had early fieldwork. Our tent pressed on the same exact rectangle of ground then. Watching the river swell with meltwater, lift an ice sheet, crack it like an eggshell and carry the shattered white acreage downstream. The vertigo of breakup.

- A breakup of a different type, this one back in Eagle, when I was a park ranger: Getting dumped by my girlfriend of five years at the same time major league baseball went on strike with both the Expos and Yankees in first place. Stepping into the British red phone booth in Eagle. Rolling in a quarter and calling my father. His unfollowed advice: "Get your ass back to Fairbanks and work it out with her."

Water under the bridge.

The Hudson of my youth is a lifetime away; somehow it has stayed with me in the Yukon's magnetic attraction. Two rivers, so different and distant, pumping water to two oceans every second I've been alive. So much water moved during my lifetime. There will be so much after.

This river is so big, so free, so wild. If you get the chance, submit to the Yukon's power. Float on its glassy back. Imagine the mass and volume beneath you.

Nose into the villages downstream from the bridge: Rampart, Tanana, Ruby, Galena, Koyukuk, Nulato and the rest. Overcome river shyness to meet people at fish camps, whose nets stretch across the river where it runs backward in mysterious eddies. Admire the tans of the river people and their ability to walk over rocks with bare feet. Their sled dogs' wolfish eyes following you from holes they've dug in the tan riverbank soil.

If you are lucky enough to make such a journey, most of your time will be this: Listening to the hiss of glacier-ground mountain against the skin of your boat, feeling the warm kiss of Interior summer sun on your skin and smelling the mint of new poplar leaves unfolding in the sun.

16

Two quotes

In the course of a long life, sometimes phrases from other people linger in your brain. They may seem disconnected, but our minds, ever seeking order, match them up.

For me, there were a few lessons on life in a note from a friend about how his life broke and a quip from a man who busted his life more than once. They have both become touchstones for me; synapses in my brain have connected them.

JOHN

Ned. Subaru broke down near the bridge. Nadia doesn't want me any more. I have seven dollars.

John Arntz, my next pipeline-hiking partner, had sent me that note from San Francisco three decades before.

My friend was describing his life after college at UAF and then completing law school in South Dakota. He had followed his girlfriend to her family home in San Francisco. Events in the Bay did not proceed as he had hoped.

Despite his circumstances near the Golden Gate Bridge, John did not flee back to the Alaska-waterless-cabin life. With no girlfriend and not much money, he stayed on in San Francisco. I'm not sure I would have made that choice.

John took a position as a union laborer, starting on the lowest rung. With his meager wages he could afford to live only in the city's Tenderloin district. There, he stepped over human excrement on the sidewalks and was nice to the junkies who shared his living space.

In the years that followed, John made a great comeback. I got to catch glimpses of it.

My observations started when my bosses began sending me to a science conference held each December in San Francisco. I loved the break from the dim Alaska winter; feeling all the sunshine there provides the cleanest flick of a switch I have ever known: Poof, I have energy again! I also have used the trip as a springboard to Outside, after San Francisco spending Christmas with my family in New York or Kristen's in Colorado.

Over the years, I have spent a half year of my life in San

Francisco. While there, I visit my cousin Heather, who has a sweet apartment on Russian Hill, and I spend precious time with John.

John now lives in a house in the Sunset District with a pastel interior he painted himself; the kitchen is tangerine, his bathroom sky blue. He welcomes me every December to sleep on his couch. I wake each morning to the squeals of playing children at the church school across the street that occupies an entire city block. Then I hop on his bike and ride through the mist of Golden Gate Park toward the conference hall.

John buys us expensive tickets to San Jose Sharks games, San Francisco 49ers games and Golden State Warriors games. Without flinching, he pulls out his credit card and pays $60 for parking at stadiums. He orders Indian food that gets delivered to his door for us as we watch sports on TV. He drives us for half a day to Death Valley, where we have executed winter backpacking trips.

Now the director of the San Francisco Department of Elections, he has come a long way in time and circumstance since he was broke and had his heart broken when he arrived there.

Unlike when John lived in the Tenderloin — or a bit later, when thugs attacked him with a baseball bat and pulled $1,000 he had tucked in pocket just after he had attended a seminar on how to feel rich — he has made it.

Due to his likability and people-managing skills, he ascended past managers in the San Francisco Department of Elections. As director, John has brought stability to a position people used to read about for the wrong reasons. For example, when unsecured ballot boxes full of uncounted votes blew into San Francisco Bay and a reporter heard about it.

With no rooting interest in politics, John is the perfect person for the job. His lack of that sort of ambition helps explain how he has stayed the department's leader for more than a decade, in a place where so many people care so deeply about voting.

I am proud of my friend's professional ascension, but that's not the only reason I love him. We get each other as only two friends of the same sex can. It has made my partners a bit jealous over the years.

"It's too bad you can't marry John," my girlfriend during my first pipeline hike in 1997 told me.

That shocked me a bit, but then I realized she was right. John is thoughtful to the point of being selfless, his house is neat and he is a great cook.

But there will be no Ned-John marriage photos appearing on Instagram. We are confirmed heterosexuals, and there is another point of common interest. We appreciate the female form together as we nod toward the running clothes of the "nonobtainables," as he calls them. (In defense of ogling while being married, I lean on the words of Jimmy Carter, who bravely spoke of the biology that affects the human animal in a 1976 Playboy interview: "I've looked on a lot of women with lust. I've committed adultery in my heart many times.")

I don't have to hide that with John. I really don't have to hide anything, which is sort of the definition of a good friend, right?

* * *

John arrived the Yukon River campsite with his right ankle swollen to nearly twice its size. He had broken it a few months before when he slipped in greasy mud during a California trail race.

"You sure that's going to be OK?" I asked, nervous that we might have to stop after a day or so.

"Think so," he said while throwing extras from the food box he had carried back into the box on our Yukon picnic table.

His words were all I needed. We have shared a lot of trail miles, another bonding point.

SUMMER OF GRAVEL AND STEEL

John and Cora

Not too long ago he traveled to Fairbanks so we could hike the Equinox Marathon Trail together. As we walked the 26.2-mile course, a cold rain misted on us all day.

The saturated course made for poor footing. On a canted dirt trail, we kept sliding off tiny ridges of mud. After one slip, John hit the ground with a thud that sounded like Rocky Balboa punching a side of beef.

"You OK?" I asked.

"Yup," he said, shoulder stained brown with mud.

As with Andy, I have shared many grinding miles with John. We sometimes run out of things to talk about. But we can both do quiet, too, without feeling a need to fill the air.

DEAN

Left to my own devices, I'm fucked.

Five miles north of the Yukon River, John, Cora and I walked up to a sign:

"BEWARE OF BEARS! Yes, it's true — they do live here. Just because you're on your vacation doesn't mean you won't get mauled! Use outhouse at your own risk."

Welcome to the Hot Spot. The Hot Spot Cafe has been one of those unique rural Alaska roadside institutions. It survived for years serving food to hungry truckers and travelers while located just five miles from the Yukon River Camp, which is open year-round and also serves food.

As we had done 20 years earlier, John and I walked up and found two plastic deck chairs at the Hot Spot. We sat down.

One of our favorite stops in 1997, the cafe is one of the most remote hamburger stands in the world. It sits on one of only five "development nodes" the Bureau of Land Management allows on the Dalton Highway. The owner, Theresa Welch

Morin, who happens to be the same age as me, leases the land from an Alaska Native who owns a small chunk of Five-Mile Camp, a gravel pad five miles north of the Yukon River.

At more than 400 miles long, the Dalton Highway has just a handful of restaurants. For years, two of those eateries were only a few miles apart. (As of 2024, the Hot Spot had opened in Fairbanks and the Dalton Highway location was closed, but maybe not forever).

On our first visit 20 years earlier, John and I met Theresa, her partner Dean Morin, and two of their boys Sean, 9 and Tyler, 7.

Theresa gave us free cheeseburgers then. "You walk 800 miles, you aren't going to pay for a hamburger."

On that day, Dean came out and entertained John and me with stories as we ate. Having the cadence of a standup comedian, he explained to us that most of the buildings at their small compound were structures they had scrounged and refurbished.

The best of those stories inspired the name of the business. The building in which Theresa cooked burgers and handed them out the window was once a bathroom with four toilets that serviced men and women working in the oil fields to the north.

Dean purchased the building for $300, removed three of the toilets and converted it for this very different purpose. (Theresa insisted that he leave one "for the ambience.") The toilets had employed electric heating coils to incinerate the waste. Thus, the Hot Spot.

"We got a perfect score on the pre-opening inspection," Dean told us. "I was elated with turning an outhouse into a restaurant and getting 100 percent on the inspection."

During that first meeting with both Theresa and Dean, John and I liked the couple right away. Theresa for her good-humored forceful nature and Dean as her quick-witted sidekick.

Their shared stories also made for my most meaningful experience of Pipeline Hike 1. Dean later told me they were both recovering addicts, something I knew a bit about. After that hike, I interviewed both of them at their south Fairbanks home and included their story in *Walking my Dog, Jane*. That interview had a big impact on me.

It was then, with my little tape recorder rolling, that Dean uttered a phrase that has stuck in my head:

"Left to my own devices, I'm fucked."

In 2017, Dean was nowhere to be seen. John and I noticed his absence but didn't ask. I'm not sure we wanted to know.

I remembered how Dean had told me 20 years earlier that addicts who paired up had about a zero percent chance of staying sober as a couple.

"Dean started using again," Theresa said unexpectedly as John and I waited for our burgers. "We are no longer together."

Then she told something I had read in the Fairbanks newspaper a few years before. Sean, the boy shooting spitballs at John through a straw when he was 9 years old and standing right over there, had as an adult killed himself with a handgun.

John and I said nothing to Theresa's somber news. It felt like everything had changed at the Hot Spot except Theresa's resiliency.

I felt I owed Theresa and Dean something. They had, like rocks in a stream, slightly altered the flow of my life.

My parents were both alcoholics — the functioning, loving kind. At times I prayed to God for my father to get pulled over when driving drunk with me sitting beside him; for the social shame that might have forced him to quit drinking. But he always got away with it, long enough to push his liver beyond function and end his life.

I drank alcohol, and enjoyed it, starting when I was a

teenager. Add a little water to Dad's vodka bottle, watch the transparent swirl inside, hope he doesn't notice.

Alcohol helped me, a shy boy, be less so. So much that at a high school dance I accepted the challenge when my friend egged me on to punch another kid (who an hour later cold-cocked me when I was looking the other way). I continued to drink every day through my twenties, through a four-year hitch in the Air Force and then the start of my independent life in Alaska.

One day when I was still a teenager, my Dad surprised me by saying I was a disciplined kid. He had watched me save money from a dishwashing job without spending it. That self-restraint is maybe what helped me later, when a few things happened that made me decide to quit drinking at age 29.

First, I had watched my parents drink for years and had seen how much of their lives revolved around filling the time before their first drink in the afternoon. Past that deadline, they were hard to talk to and unable to engage in anything except watching TV. I could see that happening to me as the years ticked by.

Second, my friend Peter from Hong Kong, who worked with me on the university newspaper staff, died one cold September night in Fairbanks. He had walked to the liquor store as he did every night and had carried home a bottle of vodka in a brown paper bag.

He drank the whole quart and then passed out, falling to the floor of his shower in his dingy apartment. His heart stopped. Pete's body lay there for hours before someone on the floor below investigated the leak.

Pete's death pushed me more in the direction I was going, having in the weeks before that limited myself to two beers a night. When he died, I dumped the beers in my fridge outside on a fluffy foot of snow that never left during the earliest winter I can remember. That was in 1992, and I haven't had a beer or any alcohol drink since, except for a sip of Champagne with my dad and brothers on New Year's Eve 1999.

Though I love beer, the transition to non-drinker was easy for me; I was mentally ready, with a bit of anger as my motivation. But I sympathize with those like my parents for whom quitting is hard or impossible. I guess I stopped when I was young enough to have the strength to do something else with my minutes. I'll never know how my life would have went had I kept drinking.

* * *

After the hike, when I interviewed them at their South Fairbanks home, Theresa and Dean inspired me to go to Alanon meetings. Alanon is for people whose lives have been touched by addiction of a family member.

I used my parents' alcohol abuse as my entry badge. At those meetings in a Korean church in south Fairbanks, I sipped coffee from a foam cup, listened to peoples' stories and shared my own. The gatherings helped me find a foothold when I felt lost after the death of my parents. Mostly they reinforced a notion Dean had told me, another of his pithy quotes:

"The good news is there is a God. The bad news is you're not it."

I was no longer left to my own devices. After a few years of those meetings, I started to take more responsibility for my imperfections. I stopped blaming my parents for all the ways I am fucked up. I realized they did the best they could, while navigating lives way harder than mine. There was a great freedom to forgiving them and deciding to beam them full love power while they were still around.

At the Hot Spot with John, after Theresa caught us up on the last 20 years, I thought of something to break the tension.

"Theresa, I have a shot of you leaning out of your window there from back in 1997," I said. "Let me find it on my iPad."

I pulled the device from my backpack. Soon, I found the picture of Theresa 20 years earlier almost to the day, leaning out the window frame that Dean had scrounged from a grader and painted red.

"Could you, maybe, pose like that again for me?"

Theresa propped her elbows on the window, bent out and smiled. I again took her picture.

"You two get together now," John said.

I wrapped my right arm around her. John then took a picture of Theresa and me, two 54-year-olds standing together in the arctic sunshine.

17

The tent

As we stopped for the night with a view of the Ray Mountains, blue and misty in the distance, John pulled from his pack a single-person green NEMO tent. That bikepacking model was as light as a loaf of bread.

Still buried in my pack was a two-person REI tent that had provided Cora and me many nights of protection from whining mosquitoes and tapping rain.

This was another change from 20 years ago. Back then, my friends and I had shared a single tent to save weight. In the two decades since, designers have shaved mass and bulk from tents by using lighter and thinner materials.

In 2017 my partners carried their own tents. What we lost in camaraderie we gained in not having to smell each other's socks.

John barely fit in his NEMO. He touched the walls whenever he rolled over. I had eyed that same style tent while in Fairbanks on my break there before heading north. But when I brought it home and set it up on a carpet inside, I found its fatal flaw. The single-person tent had so little room that Cora refused to enter, no matter how nicely I encouraged her. I didn't blame her, as my torso and legs took up most of the floor space. It felt like I was inside a dip-net.

I returned that tent to the sporting-goods store. The two-person model I bought in its place weighs twice as much but is still less than five pounds. It has ample room for me to lay with my journal on the floor and write while leaning propped on an elbow. Down at my shins, Cora has plenty of room to stretch on her little foam pad.

Some nights, especially the buggy and rainy ones, I lay on my back and stared at the elegant web of poles supporting the netting and the rainfly. As the frustrated mosquitoes and gnats swarmed the screen, I smiled. That ever-so-thin barrier was the difference between a good night's sleep and never-ending harassment.

And when it rained, instead of drops on my face, I fell asleep easy to the rhythm on the rainfly. After a few drizzly mornings, I became an expert on just how much water was coming down by the frequency of taps. The rain was never as bad as the drops on the fly sounded.

The tents I carried in 1997 were also good, lighter than tents from 20 years before that, but a bit heavier than the 2017 models. Tent materials and designs continue to improve and adapt, offering us more options, usually for less weight and more money. We have come a long way from an animal skin stretched over a frame of wooden poles, and from the canvas family tent my Dad set up every 4th of July in upstate New York. The box for that tent could have been used as a cardboard coffin. I loved sleeping in it, and the giant, beefy metal zippers that are probably still intact somewhere in the Hudson Falls landfill.

* * *

After a few dozen nights, I got to be pretty quick at finding a flat bed of lichen or moss with enough of a rectangle to accommodate the tent. After throwing down the clear plastic ground cloth, I would sometimes lay on it to see if my feet where lower than my head. I wished for a carpenter's level every time, wondering if my estimation of plumb was correct.

When I broke camp, I tried to remember to stand over the pressed-down grass and bow, giving thanks for another safe night on the ground. I also took a picture of the tent site.

* * *

The fabric of my current tent is breathtakingly delicate,

not much thicker than tissue paper and supple as a bat's wing. The teensy zippers seem like they belong on a windbreaker. To save weight on my pipeline hike, I left behind the tent's stuff sack. I jammed the tent and fly fabric straight into my pack, where it molded around my compressed sleeping bag and rolled-up cylinder of air mattress. That fit the shape of the pack better, with less wasted airspace. I held my breath while shoving the delicate rainfly in, but it survived each cramming.

To make my little home more livable, I cleaned the interior each day of the lichen, dirt, and dog hair. But rather than do what I had in the past — hold the still-erected tent upside-down over my head and shake it, raining debris on my face — I started cleaning it a different way.

After removing the rain fly, I left the tent staked out and released the plastic clips from the poles. The tent deflated into flatness.

Leaving one door unzipped, I pulled the stakes, still cool from the ground. I lifted the flat tent body from the ground through the open door, pulling it inside out like a sock.

Turning my back to the breeze, I shook the reversed tent like a sheet. Out flew all the stuff that had hitched a ride in with Cora and me. Even from the corners. I then pulled the insides back and through the open door. I zipped the door shut and stuffed the fabric into the bottom of my pack.

This new inside-out method worked so well I wondered why I hadn't thought of it years earlier. Not only does it keep spruce seeds out of my eyes, it leaves the wind no chance of carrying a tent off like a kite.

* * *

I don't fall in love with many material things, but my tent is an exception. Knowing I have that sanctuary after a long day of hiking makes me happy. That respite is always there on my back when it seems the bugs are unescapable.

On some nights, I remembered to be thankful for the shelter that rode on my back, another miracle of our time.

Night 61, July 5th, Prospect Creek:

Where it is 80 above rather than 80 below! (North America's all-time low temperature, measured on a thermometer here on Jan. 23, 1971).

My bud John left me a mile back, as the blue Northern Alaska Tour Company van scheduled to arrive at the nearby pipeline access road at 4:45 p.m. did so. I had a good time answering questions for the tourists within, especially a family from India on their first trip to Alaska.

I am missing John as I pitched the tent in the Prospect

Creek lowlands, surviving bugs only because it's so hot. I didn't want to walk past Pump Station 5 until tomorrow (because I'm feeling bushy and shy), and the pump station is a ten-minute walk from here.

After he bathed naked in Prospect Creek, John stepped up into that blue van. I got teary at being alone and missing a friend who really appreciates this, the robins and the sculpted green hills and valleys thick with spruce. He's a good man and 20 years of friendship is no small thing.

Sun is still a big hot ball at 9 p.m. But Cora and I are already in the tent on crunchy lichen, shuttin' 'er down. Alone for the first time since Tolovana River summer solstice, two weeks ago . . .

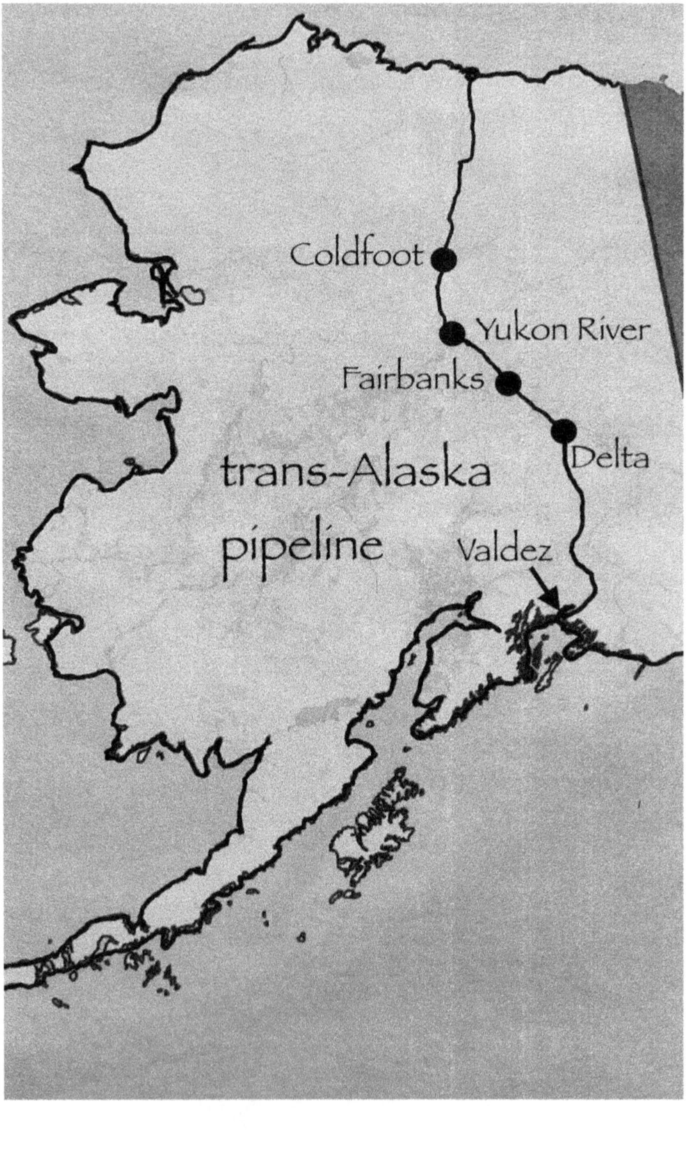

Coldfoot
Yukon River
Fairbanks
Delta
trans-Alaska pipeline
Valdez

18

Northern cat

Even with the sudden, expansive view to the northwest, I didn't realize that Cora and I had walked into the basin of one of Alaska's great waterways. It took a few hours until it came to me that the Koyukuk River had cut that rift in the smoky blue hills.

The Koyukuk is born in the country of tilting limestone mountains to the north. Those hills are the final of three broad spines of rock we would walk up to, and then over. The Brooks Range.

A few days after we entered the basin, the pipeline's path led us right to the shore of the Koyukuk. There, aquamarine water cleansed open gravel bars. Skipping stones aproned the islands. Willow whips waved in the breeze. I imagined a river trip and wanted my red canoe.

Cora and I stopped on the river bank — another subtle-yet-meaningful marker showing our slow progress across Alaska. She lapped at the green-blue water. I boiled a bit of the river for a coffee and ramen. When the coffee was ready, I cupped my steaming red mug, looked at the river and thought about where all that gorgeous liquid was heading.

The Koyukuk starts by trickling clear from stream cuts of the whitish mountains. It then snakes its way slowly — almost like a backwater slough in the lower reaches — for 425 miles to the Yukon River at the village of Koyukuk, just downriver of Galena.

A comparable Lower 48 river by size and the area drained is the Susquehanna, which flows from near the Baseball Hall of Fame in Cooperstown, New York, to Chesapeake Bay in Maryland. Along the way, that river catchment receives about half the moving water in Pennsylvania and then carries it to the Atlantic.

The Susquehanna drinks in groundwater, snowmelt and rain not soaked up by sugar maples, hickories, and beech of the temporal deciduous forest. The Koyukuk's dominant trees exhaling vapor to the atmosphere are bank willows, balsam poplar, aspen, birch and the spruce — black and white — of the boreal forest.

Cora and I were stepping closer to the northern limit of that great band of trees, which sweeps from northwest Alaska to eastern Canada. We were within two weeks of outwalking the forest.

A few thousand miles closer to the equator, the brushy and cleared banks of the Susquehanna have been way noisier than the gravel we were crunching across. This busyness included, once, the construction of a nuclear-power facility on Three-Mile-Island, and all the fuss it caused when a reactor partially melted down in the 1970s.

The Koyukuk's muskeg and grass shorelines have remained much less dramatic, except during floods. In modern times, the largest of eight settlements along its serpentine twists is the village of Huslia. There, about 300 people live a few hundred miles downstream of where I sipped my river ramen and coffee.

In total, 1,000 people live along the big, gentle Koyukuk. That compares to 3.8 million souls living vastly different lives along the Susquehanna, with many of those people unaware of the ever-moving conveyor of water nearby. Once again, the Alaska Difference.

* * *

In the woods near our campsite on the middle fork of the Koyukuk, Cora and I made a discovery: Rectangular metal cages about the size of a kennel for a Chihuahua. The traps stood with doors open, unset. An identifying tag showed they belonged to scientists with UAF's Institute of Arctic Biology and the National Park Service.

Nearby, we soon found a homemade chicken-wire and plastic-pipe cage large enough to capture Cora. I knew whose traps they were.

Those non-lethal animal catchers belonged to Knut Kielland, a Norwegian ex-pat. With these traps, he and Donna DiFolco, of the National Park Service's Fairbanks office, have studied snowshoe hares and lynx for years.

Knut uses folksy expressions that have cracked me up when quoting him for stories over the years. He once referred to spruce trees he studied as "pumpkins." When a few dogs passed, he also called them pumpkins.

Knut is tall and wears glasses, mukluks and a baseball hat with a tiny bow on the brim. That hat is favored by old-time rivermen of Alaska's Interior. Knut has a kinship with those men, for all his time mushing and snowmachining and studying the Tanana River downstream of Fairbanks. He smells like spruce smoke.

Knut has invited me several times to the Tanana just downstream of where the Chena empties into it, giving me new appreciation for the floodplain willows and their importance to hares and moose and so many other creatures.

I once visited Knut in the Koyukuk country in March. That time, I was pretty excited that Knut had asked me if I wanted to come up for a week to Marion Creek north of Coldfoot to help him with his lynx study. His invite coincided with my birthday. What better way to celebrate another full spin than to maybe get close to the mysterious cat of the north?

One cold morning on that trip when I accompanied Knut, he trapped a lynx. Following is a story that I wrote for the Alaska Science Forum about that experience. I have changed it a bit from how it appeared in newspapers.

* * *

NORTH OF COLDFOOT — The lynx looked out from inside a chicken-wire cage. Despite its loss of freedom and the nearby squeaking of boots on cold snow, the wild cat looked calm, as if it might be cooking a snowshoe hare in its belly.

Knut Kielland, a professor with the University of Alaska Fairbanks's Department of Biology and Wildlife, years ago would trap lynx for their fur. But now he captured this 22-pound female lynx as part of an Alaska-wide project he leads to better understand the ecology of the cat that has wandered farthest from the equator. North of Coldfoot, there is not much farther to walk for these relatives of lions, leopards and cougars.

Knut and the lynx

Knut started live-trapping lynx back in 2008 at the Bonanza Creek Long Term Ecological Program in the forest near Fairbanks. Here north of the Arctic Circle, he partnered with Donna DiFolco of the National Park Service in 2014 to study lynx.

Expanding what Knut started, biologists at both Tetlin and Yukon Flats national wildlife refuges are also trapping and following lynx, as have scientists with Kanuti and Koyukuk-Nowitna refuges.

In hushed tones, Knut asked for help in preparing a syringe attached to a pole. He measured out a dose of anesthesia that would immobilize the lynx for one hour.

While Knut worked in the 5-below-zero air, the lynx stared out, not blinking. This was maybe the same animal that visited this same trap one week earlier, after Knut wired the guillotine-style door open with bait inside. His trail cam had snapped a photo as a lynx stepped in to get it.

With the local snowshoe hare population at a low in its cycle and its primary food harder to come by, this lynx returned to the trap in which Knut hung a bloody roadkill grouse. Knut had stopped for it on his drive up the Dalton Highway from Fairbanks, opening the back door and setting the dripping prize on a piece of cardboard.

Sniffing out that treat, the female lynx stepped inside the trap and pressed a broad paw on a rectangle of plywood. That action stretched a wire that pulled a bolt from the trap door, which fell shut the night before we visited.

Knut approached the trap with his syringe attached to the 4-foot metal rod. The cat didn't move.

Through the chicken wire, Knut poked the lynx in a rear haunch. It jumped, and then settled down. Its tongue darted in and out of its mouth. Knut knew this serpentine reaction meant the drug was working.

After a couple of minutes of silence, Knut whispered.

"Go up and tell me what you see."

I squeaked a few steps toward the trap on the cold snow.

"Looks like it's sleeping," I said.

Knut waited another minute. He approached, and then poked the lynx with a spruce branch again through the chicken wire. The animal, a pile of lushness, did not react.

Knut then opened the trap door and secured it with the bolt. He reached in and gently pulled out the lynx, much as he once lifted his two sleeping daughters off the couch when they were young. He laid the limp cat on a foam camping pad on top of the cage.

In the cold March air, I pulled my hand from a mitten. This was a rare opportunity to touch a wild living creature. I placed my hand on the small of the lynx's back. I felt the steady and slow rhythm of its breath. I didn't detect any heat through its thick, supple fur until I pressed down a few inches.

I touched my hand to its paw. Oh my. The pad, puffy with hair, made my hand disappear beneath it. Snowshoes, designed by nature. Its mighty claws were retracted. I felt for them behind the lynx's thick foot pads, but couldn't detect any sharp points.

The trapper/ecologist who lost his taste for killing things like lynx has handled many live animals. Knut worked fast and gentle, which seems to me a debt you owe a wild animal in exchange for information.

Knut lifted the cat's rear leg and squinted, determining the lynx was a female (he had initially guessed male, because of her large head and lanky body). With bare fingers, he felt the animal's sides and belly.

"She's pretty lean," he said.

He read the animal's pulse and blood-oxygen level with an instrument attached to a plastic clip that pinched the lynx's tongue. He measured the length of its hind foot and its ear tufts.

He adjusted the plastic collar to the thickness of the lynx's neck by matching up holes in the strap. He secured the two collar bolts with a nut-driver. The block of transmitter and battery on the collar would allow him to see where the animal wandered via satellite. The device also has a timer: the collar would pop off the animal late that summer and drop to the forest floor.

As the lynx began to twitch, Knut performed the final task — installing a small metal ear tag with a specialized pair of pliers.

As Knut squeezed, the lynx suddenly looked up with blazing yellow eyes; a chill ran through me. I imagined that piercing look as the last thing a snowshoe hare might see.

With the science completed, Knut returned the lynx gently to the cage. He wanted it to recover from the drug before he set it free. The anesthesia was wearing off.

Soon, the lynx scratched at the chicken wire with claws, which had suddenly curved out like daggers from a sheath. Knut lifted the door. The lynx scampered out. It stumbled toward the frozen creek, disappearing into the spruce forest.

"I sure hope she stays around," Knut said.

The ideal scenario for him and his fellow researchers was that this female would mate and give birth to kittens here in the valley of the Dietrich River. But several of the lynx collared here — the first lynx in Alaska ever fitted with satellite collars — took off.

One wandered to Cape Krusenstern on the Chukchi Sea north of Kotzebue, a place with tundra vegetation that does not nourish snowshoe hares.

Another, nicknamed Goldie, roamed from here to Norton Sound, passing near the villages of Koyuk and Shaktoolik before continuing to the mouth of the Yukon River.

A third went all the way to British Columbia in Canada, a distance of 1,000 miles.

"One thing we learned is their capacities for dispersal," Knut said. "Not only the sheer distance, but straight across country, sometimes almost on a compass bearing."

Why do lynx take off, even to where there are no trees (or their favorite food, snowshoe hares)?

Scientists don't really know that answer yet, but they suspect that hunger is a major driver.

Before his three weeks was over there in the Brooks Range north of the Arctic Circle, Knut would spend many hours baiting, setting and checking his 20 live-traps along a 56-mile line he accessed by truck and foot along the Dalton Highway. He would capture and collar one more lynx, a male Knut and his helpers saw on trail-camera images last fall.

He returned home and to his office at the UAF Institute of Arctic Biology. There, he logged into his computer and saw where the recently trapped female lynx, the one Knut let me hold, had been roaming.

Knut named the lynx Jenna. Jenna is also the name of Donna DiFolco's now 20-something daughter who, as a little girl, spent much time in the Brooks Range with her mother setting traps for hares and counting piles of hare pellets.

As the months went on from that April day when Knut captured and collared Jenna, the transmitter around her neck sent blips thousands of miles into space. An orbiting satellite registered where the 22-pound cat was roaming as the cold spring turned to melt season.

Until Jenna's collar popped off during the endless light of summer in late July (a bit earlier than Knut had programmed), she proved herself a bit of a homebody.

Her travels were all up and down the main drainage of the nearby Dietrich River and the clear side creeks, including the one where she was captured. The farthest tracking point was just 20 miles from that trap.

Unlike several female lynx Knut's team had captured in the past, Jenna's dots did not stop moving in early summer. When others had, the scientists knew the lynx had expanded a hole beneath an uprooted tree or found another secluded spot. There, they gave birth to kittens.

Jenna was not a mother that spring, which was perhaps due to few snowshoe hares hopping around.

Hares follow a mysterious 11-or-so year cycle of boom and bust. Lynx and the larger boreal-forest animals that depend on snowshoe hares — including owls, foxes, coyotes and wolves — will track the fate of those three-pound bundles of stringy protein and a touch of fat.

But the hares that were once hungry enough to girdle bitter spruce saplings were no longer so numerous in the upper Koyukuk. Born a little too late, Jenna will probably be lucky to live a full life in this hungry country.

* * *

Our Koyukuk River campsite was on a clearwater channel of slough-like water stranded by the river's drop after the pulse of spring snowmelt. It smelled minty fresh, like balsam poplar leaves.

The camp featured a nice sandy beach that begged to be walked with bare feet. Doing just that in the morning, I saw Cora sniffing some tracks in soft mud by the water. A closer examination showed perfect, circular cat tracks.

The lynx had passed while we were sleeping in the tent. I looked to the willow brush where the tracks led. Sometimes, lynx will sit and stare at people and even their dogs. In those moments, they seem to have no fear. I squinted at the willows and wondered if a pair of blazing yellow eyes might be peering back at us.

19

Trail angels

One of the many ways in which I have been lucky to be alive now: This period has coincided with a Golden Era of global shipping. Especially when it comes to food.

The Alaska Myth is that of the Northern Serengeti, a place teeming with charismatic megafauna visible with little effort. More caribou than people live here, it's true, but they can be hard to find when you want to eat one. The problem: All those creatures are scattered over 665,400 square miles.

The hares of the upper Koyukuk River valley were not zipping across Cora's and my path very often. Nor were grouse, voles nor anything else that might nourish a person.

On the hike 20 years ago, hiking partner John Arntz fished for and caught a grayling in Douglas Creek. A pipeline security guard had given me a spinner and some fishing line that

John tied to a stick. That wriggling fish represented the only time I had gathered a meal from the land during both journeys. One six-inch grayling wrapped in foil and cooked on willow coals was not much of a dinner for two hungry boys.

Most all of my food came sealed in plastic, from processing buildings in California, Oregon, Michigan. A few originated in Alaska, including Heather's Choice smoked-sockeye dehydrated dinners. Cora's dog food came from Outside by barge and truck. I had divvied those factory-compressed nuggets from 45-pound bags into gallon Ziplocs and then thrown those into food-drop boxes.

Except for a few overwintered-yet-intact lowbush cranberries — the tartness of which fired my salivary glands until they ached — I was not tasting much of Alaska on my long walk. And I have not gathered much food from the big land during my half-lifetime here.

It takes skill to harvest fish or mammals. Dropping a moose in the woods is like suddenly needing to part out a grand piano for the backhaul to a road. And it is amazing that more of us don't perish on the highway after the stupefying process of scooping 30 five-pound fish out of the Copper River in 15 hours of sweaty effort, then cleaning them on a metal table, hefting coolers into the car, shooing away the gulls at O'Brien Creek to rinse your hands one last time and turning the key for the 8-hour drive home.

On every step of the trip across Alaska, Cora and I were dependent on food from far away. We couldn't carry 100 days of food on our backs.

* * *

Michelle: Trail Angel.

Hi Michelle. Need box No.8 at Coldfoot July 23 or close to then. Doable?

A few hours later after I typed that message into my inReach satellite communicator, a pinpoint of green flashed on my inReach. Michelle's reply took an instant to travel 44,000 miles: From Fairbanks — less than 300 miles away on the back of a raven — to a geosynchronous satellite zipping through frictionless black space and back down to Earth and me, a guy scuffing along a gravel road next to a giant pipe north of the Arctic Circle.

Roxanne is driving a group to Coldfoot on the 13th. She will bring your box that day!

* * *

Michelle Charlton is a supervisor with a tour company that provides visitors with trips in vans from Fairbanks to the Arctic Circle, the small settlements of Coldfoot and

Wiseman, and sometimes all the way to the Arctic Ocean, a 450-mile one-way trip executed over a few days.

I found Michelle through a mutual friend who had driven one of those vans as a summer job. Before the hike, my friend suggested I contact Michelle to consider delivering the 11 food boxes I had packed for the trip from Fairbanks northward. With so few communities and almost no postal service, I needed some specialized help.

"I love Walking my dog, Jane," Michelle emailed back the same day. "Of course, we will help you."

I arranged the details while in Fairbanks. During the hike, I would text Michelle whenever my pack was getting light enough for another resupply. Despite her son being a toddler at the time, Michelle always answered me within hours, on weekends and on her vacation days. Then she would send one of her people to my house and the unlocked guest cabin a few dozen steps down a plankwood path.

The cabin held my late-summer stash of food in 11 stacked cardboard boxes, each labeled in Sharpie with an encircled number.

Michelle's friend would find the box I had asked for and carry it to the tourist van at the airport. Soon, the box would be rolling on its way to me.

Michelle did not ask for anything in return. I told her, though, that it would be fun to talk to the tourists who were riding in the van while I stuffed the new food in my and Cora's packs.

"That would be great," Michelle said.

It was. The passengers would pile out of the van on a gravel pull-off somewhere close to the pipeline but away from truck traffic. The driver would introduce me to the dozen visitors. As they swiped at mosquitoes, I would describe what the last few days were like and how many moose or bears I had seen.

Cora would make the rounds, weaving in and out of people, getting her back dusty as she offered her belly for pets. People from India and Indiana would ask me questions about where I camped every night and why did I want to do this. We would always find something in common between us.

I loved the distraction, especially when hiking the pipeline seemed dumb. While walking that gravel road next to a huge pipe that led me past valleys and rivers leading to wilder adventures, I sometimes wondered what the heck I was doing. But those people made it seem like I was doing something interesting. And, like my dad the high-school teacher, I love to perform.

Jillian: Trail Angel.

At the Coldfoot truck stop, a quarter mile from the middle fork of the Koyukuk, the good folk kept appearing.

One day, Cora and I walked onto the deck of the Coldfoot Cafe, wet and dripping from a thunderstorm.

With round droplets still falling from the bill of my baseball cap, Jillian Simpson appeared on the deck with a smile. Blond and striking, Jillian had introduced herself to me a few weeks before when I was eating a burger at Yukon River Camp, owned by the same tour company for which Michelle works.

Three Fairbanks high-school buddies had created the tour company years ago, and now they own many structures, an airline and some fantastic real estate in northern Alaska. Jillian is married to one of those pals, who flips burgers and takes out trash with his workers here in Coldfoot.

"You can have a room here tonight (for free)," Jillian said. "And you can sneak Cora in. Just try to be quiet."

She then invited me into the home she shared with her husband and two young daughters that was 100 steps away from the cafe.

"You can do some wash here, and do some writing. I can give you a hotspot from my phone for internet if you need it."

I walked Cora inside the house as Jillian returned to the restaurant. I sunk into a couch and couldn't help but smile. Twenty years earlier, the manager of this truck stop led me into this very same house. Is the washer into which I would stuff my clothes the same one I used back then?

* * *

John: Trail Angel

While fattening up on a bacon cheeseburger at the Coldfoot Cafe a few hours later, I looked at my map. I wanted to walk into the town of Wiseman, about 12 miles away, but it was not on the direct path of the pipeline and was on the wrong side of the Koyukuk River. To get there, Cora and I would need to do as Jane and I had done 20 years earlier: walk past Wiseman to the Dalton Highway bridge over the Koyukuk, then backtrack down a three-mile side road into Wiseman.

On the map I saw another way. Before I got to that bridge, if I detoured from the pipeline down the bed of Minnie Creek to its mouth, I would be standing right across the Koyukuk River from Wiseman. The river was too deep there to wade, but I could get across with a boat. That would save my sneaker soles about 10 miles.

The Alpacka company had sponsored me with a packraft. I had used it several times before, but was not carrying it there in northern Alaska to save weight.

Later that same long evening, Cora and I walked over to the Bureau of Land Management Interagency Visitor Center, a slick, beautiful building with a theater that seems out of place in Coldfoot.

With Cora roaming the theater, I gave a presentation for the visitor center staff after they asked me to. When I finished, I spoke with a longtime Alaska ranger named John. I told him I wished for a boat so that I could ferry across the river to Wiseman.

"Why don't you take our packraft?" John said. "It's drying now on the porch because I just cleaned it. You can just leave it with someone in Wiseman and I'll pick it up later."

* * *

I lurched out of Coldfoot the next morning with an extra-heavy pack that included the raft, rolled into a loaf in the bottom of the main compartment, and a paddle, separated into two pieces stuffed into the back mesh.

Cora and I busted out a 12-mile day in the 80-degree heat, heading up the pipeline and then a short way down the bed of Minnie Creek. I dropped my load on an exquisite gravel-and-willow island. Around us flowed clear, ankle-deep water. Cora and I fell asleep to its gurgle.

In the morning, I woke thinking we could be deep in the wild rather than a quarter mile from a highway upon which roll more big rigs than maybe any other road in Alaska.

I sipped three Vias from my red plastic mug and performed some gravel-bar yoga before getting ready to move. After breaking camp, Cora and I walked down the stream and reached the middle fork of the Koyukuk. I sat down on its banks and blew up the packraft.

With my backpack and my dog for ballast, I ferried across the smooth river. In less than two minutes I bumped the nose of the boat into riprap boulders on the far shore. Cora jumped out.

There, I deflated the raft beneath a weathered plywood sign at the entrance to Wiseman:

Wiseman was established in 1908 and has been a viable town ever since. All the cabins and property here are privately owned so please respect people's belongings.

Though friendly enough, that sign had made me feel self-conscious on my first trip to Wiseman 20 years earlier. But I had met people in the log-cabin community then, and after 20 years my middle-Alaska connections had tentacled to the extent that I knew most of the people I was to meet in Wiseman.

A partial list: Harlow Robinson, Jay Cable and Tom Moran (all there to run a wilderness race — Jay had a bag of Cora snacks from daughter Molly). Wiseman residents Jack Reakoff and Bernie and Uta and their son Leo Hicker. Knut Kielland in to do more work on his lynx/hare studies. John Shaake, a visitor from Fairbanks whom I had interviewed a few times about the Chena Lakes Flood Control Project south of Fairbanks. And, of course, there was Clutch.

* * *

Clutch: Trail Angel.

Carrying the rolled-up wet packraft under my left arm, I walked the flour-soil road that led to the heart of Wiseman. There was an appealing acre of mowed lawn that opened to the aquamarine river.

Cora and I approached a frame building with green siding and a sign tacked near the door: *Wiseman Gold Rush B+B Office. Sourdough Clutch.*

Right on cue, out of the door stepped Clutch Lounsbury.

Clutch resembled Santa Claus as he might appear on a hot summer day: tan Carhartt overalls, a white Fairbanks Hockey

Hall of Fame t-shirt stretched to its limits, glasses tinted black in the sunlight and a CAT baseball cap.

"Well, hi, Ned. We've been waiting for ya," he said, offering a meaty hand for a shake.

The son of Fairbanks gold miners, Clutch and his brother George had both purchased lots in Wiseman after meeting the old-timers there. The brothers 20 years ago had day jobs — Clutch drove plow trucks and graders out of the Coldfoot Department of Transportation compound — but they both also worked mining claims not far from Wiseman.

On my first pipeline hike 20 years ago, the pair carried me and my dog Jane in a pickup truck to their claims. We got to see what real earth-moving looked like.

The two have long since retired. George had sold his riverside cabin to the tour company and was spending his summers as well as his winters in Fairbanks. Clutch maintained a summer presence in Wiseman with the group of tidy, small cabins he rented out. The buildings were surrounded by vintage mining equipment he had gathered over the years, including a bathtub-size, sheet-metal urinal behind a cabin that was still operational.

Clutch grilled moose burgers for me as he told stories of the July 4th party he hosts every year in Wiseman; the latest had been a few weeks earlier.

There was no dead air. Clutch talked so much that I didn't have to, reminding me of a Fairbanks friend who was in his late 80s. I once compared that friend to a pipeline check valve, which allows the free passage of oil in only one direction. Many men, I have observed, morph into check valves as we age. Sitting with us gives you practice in being a good listener.

Clutch said I could camp on his lawn. I told him my wife, daughter, and cousin were arriving on the tour van later that afternoon.

"They can stay in my cabins," Clutch said. "I'm happy to have them."

* * *

The girls — Kristen, 10-year-old Anna, and Heather Liston from San Francisco — rolled in on later that evening.

Within an hour, Clutch was teaching Anna and Heather how to pan for gold in a washtub he had set up nearby. When I offered him cash for the two cabins in which he was letting the girls bunk, he waved me off.

SUMMER OF GRAVEL AND STEEL

Clutch, Heather and Anna

I wrapped my arms around his girth and gave him a hug. He seemed a little embarrassed at first, looking the other way with a pained squint. But then he squeezed me back so hard my back popped.

A warm rush flooded through me, a feeling of appreciation for those who were helping me along the trail and asking for nothing in return. Why would they do that? That was a hard one for me to comprehend.

When I was in the Air Force, 18-year-old me noticed that a guy who raised his hand to volunteer for an undeclared task was soon scrubbing pigeon poop off the parade grounds in 100-degree Texas heat. My hand never went up after that.

But I also recognized the feeling that came from volunteering to teach ski lessons to kids, when they come up to you with a thank-you note and a coffee card after the season is over. That gratitude I felt for Clutch, Michelle, John and Jillian also, maybe, made them feel good.

* * *

Later that evening, I asked Kristen how much she had paid for the five-hour ride in the tour van. I imagined three seats on that van could have gone for as much as $1,000.

"Michelle didn't charge us," she said.

20

Girls

Girls, girls, girls.

I have spent much of my time on Earth with them. Mom, until she died in 2005. Two sisters, one older and one younger, until I left home to join the Air Force. My wife and daughter for the last 15 and 10 years. Each of my four bosses during my long pull as a science writer. Both of my pipeline dogs were females too.

Though they often perplex me in ways I don't understand, I appreciate the presence of women. They look at things differently than most men, and they take up less room. When I sit next to them on the bench during a hockey game, a few of them smell flowery, in pleasant contrast to some guys whose pads reek of ammonia.

I spent a bit of time with women on my 2017 pipeline hike,

including Heather, my cousin who lives in San Francisco. She came north, along with Kristen and Anna, to hike with me for a few days.

Heather brought a splash of color to the grayish gravel bar of the Hammond River. She wore a floppy red sun hat and a red tank top. Even her backpack was red. I remembered her red couch in her San Francisco apartment, as well as her closet full of red dress shoes.

Her fondness for that color, also my favorite, did not indicate a lack of attention to backpacking essentials. Her screen-mesh tent, pitched to almost touch ours, matched the rest of her kit for its impressive lightness and compactness.

A year older than me, Heather has a magnificent explosion of reddish blonde hair and the high cheekbones of the Irish grandmother she and I both adored.

Heather acts as historian for my mother's side of the family, sending us photos she has found and preserved, helping to keep the Irish Listons who passed through Ellis Island from being forgotten.

Also in her apartment is a picture of my mother as a 12-year-old girl. Looking back at me with my own face, young Mary stands next to Heather's late father Bill Liston, then a boy. Seeing Mom there stops me for a few seconds each time I visit.

Heather's life has been fun to follow. She works as a financial consultant/tax expert. A hiker and runner, Heather once invited Anna and me to run San Francisco's Bay to Breakers 12-kilometer footrace. Some participants wear costumes, a few of which are skimpier than others. One of my enduring memories includes my daughter and Heather running past a man's dimpled naked butt.

While traveling to our first-night camp on a bar of the Hammond River, Heather hiked next to Anna, who wore synthetic hiking boots that seemed large for her moose-calf legs, which seemed to grow millimeters with each day. Anna does not freely engage with adults, but Heather pushed gently.

"It looks like Cora's been lifting weights," Heather said to Anna as they walked along. "Do you think she is carrying dumbbells in her pack?"

"No," Anna smiled at the silliness.

Many adults can't seem to find much to talk about with kids. But some — like Heather and my friend Jay Cable — seek opportunities to engage. Maybe they prefer kids' innocence.

While we hiked, Anna was chatty with Heather and her mom and sometimes even me. She comes alive when someone draws her out with conversation and listens to her answers.

Photo by Heather Liston

At that campsite, the trans-Alaska pipeline, a constant throughout most of the trip, was nowhere in sight. We had seen it just before, while following an access road that bisected the gravel pipeline pad. There, we crossed beneath a goalpost-like "headache bar" that prevents over-tall trucks from hitting the pipeline. We then hiked beneath the slighter higher elevated pipe and followed the access road to a blank gravel space.

That spread was once the home of Dietrich Camp, where a few hundred people slept and showered and ate dinner while making small fortunes in the early 1970s.

The road stopped abruptly at a 10-foot cliff overlooking the Dietrich River. That wall had formed when the river had meandered eastward to eat parts of the road and a gravel pad that bulldozer drivers had leveled to host the trailers and a runway long enough for jets.

From there we could see no portion of the pipe. And the flat lay far enough from the highway that we rarely heard trucks rumbling north and south. We could see nothing man-made except the access road.

Dietrich Camp

What remained: A sweeping view of pyramidal Mount Dillon off one shoulder and the limestone shark's fin of Sukakpak Mountain off the other; 20 acres of flat stones sorted by the river over the years into a smooth fan; and a magnificent open space that allowed a spruce-scented breeze to flow over the warm gravel, pushing the bugs away.

SUMMER OF GRAVEL AND STEEL

The sun blessed us on this day, warm enough that each of us, except 10-year-old Anna, stripped down for a bath in the aquamarine river 100 steps from our tents. Seeing my California cousin splashing in an Arctic river made me smile.

With a little imagining, the campsite made me feel as if we were on a river trip, days from any road. The only thing missing was an overturned red canoe pulled up on the riverbank, its bow tied to a willow.

* * *

"Listen, we've got to talk."

Kristen had pulled me aside at the sublime campsite, after I had criticized Anna for getting sand in the tent. I could tell by her eyes it would not be a happy conversation.

"You've got to stop being an asshole," she said. "You are so bushy after being out here with Cora for so long. I know you've got your system, but you've got to back off. Or we are going to catch a ride and go home."

I had become so anally perfect in my camping routine that I had to comment on any imperfection I witnessed.

That extreme critical eye and the need to voice it, a gift from my dad, has made me an excellent editor, but it is not so good in other life situations. Kristen reminds me every now and again, as she did there on the pipeline pad when Heather and Anna were skipping stones in the Dietrich River.

I didn't like to admit it, but I knew Kristen was right. I decided there to lighten up a bit. We had both compromised like this many times in our 15 years together. Her willingness to call me on my shit is part of the glue that has kept us a unit.

Another thing you should know about my wife: Every dog we have invited into our lives absolutely adores Kristen. They wait for her to rise in the morning, then whine and sink their teeth into her pajama leg as she drags them to the bathroom. Two of them once hurt her by knocking a vacuum cleaner into her as they jockeyed to be within her aura.

No matter that I have steadfastly escorted those dogs on daily maintenance loops spending hours longer with them than Kristen; they always prefer her. That character endorsement from the purest of spirits says a great deal about Kristen. It also makes me jealous.

"Animals just like me," she has said with a shrug.

SUMMER OF GRAVEL AND STEEL

Kristen

But there was once an animal that was not so nice to Kristen.

Besides Heather, Anna, and Kristen, another companion joined me on this section of my hike: A shotgun with a stainless-steel barrel and stock of black plastic. The gun had passed through this country before. I had carried it the whole 800 miles across Alaska 20 years earlier.

On this trip, I had lugged it along it only twice, both times when Anna and Kristen were with me. Anna didn't seem to have an opinion on the gun, but Kristen insisted the weapon come along.

The reason can be found within an Alaska Magazine column I wrote during the decade I was the magazine's "Wilderness Adventurer."

I met Kristen at a running race. A former girlfriend of mine introduced us, and I learned that Kristen was a bird biologist who had spent the previous summer working in the forest near the McCarthy Road. I thought back to the radio report I'd heard while I was walking the pipeline.

"Did you know that biologist who was attacked by the bear out there?" I asked.

"That was me," she said, quietly, before changing the subject.

She's such a lighthearted person, with a smile being the default mode to which her face always returns, that it's easy to forget that she has undergone such a trauma dealt by a wild animal.

Her fight to overcome her phobia of camping has been a success; there now seems to be little residue of fear when she rolls out her sleeping bag for the night. You couldn't blame someone like her if she never ventured near wild places again. But Kristen jumped right in.

And she has the best bear story I've ever heard someone tell, but she only gives it up when prompted.

The last time she told the bear story was one of the most memorable. She was in the hospital in Fairbanks and had been up all night with teeth-grinding contractions so painful she couldn't speak. After more than 12 hours of a new mother's agony, she nodded with vigor to the doctor when he suggested an epidural, anesthesia that would numb her from the waist down.

She endured another excruciating hour of contractions before the anesthesiologist, who was busy with a C-section in another room, arrived. Practiced in his craft, the anesthesiologist distracted her with small talk about things other than childbirth when he noticed something to ask her about.

"Where'd you get these scars on your shoulder and back?"

Just then, the drug kicked in, and Kristen regained the ability to speak. There, in a hospital room, at four a.m. on a Tuesday, she told her story.

The anesthesiologist's eyes widened as he heard of a charging grizzly that didn't stop after absorbing a muzzle full of pepper spray. The doctor and medical student observing at bedside weren't so sleepy anymore when they heard about how the bear knocked Kristen down, clawed her back and the rear of her head, and bit through her wristwatch. And her wrist.

They stood there at that early hour as Kristen described how she tried to crawl away after the initial attack, then felt the bear whack her back down to the ground. That's when, using its jaws, it pinned her head to the leaves, and she thought her time had come.

In the quiet of the Women's Center at Fairbanks Memorial Hospital, the medical folks listened as Kristen described how the bear eventually left her alone, and how after what felt like an eternity she crawled and walked out to the McCarthy Road, where her coworker picked her up and drove her to Glennallen. There, doctors cleaned and stitched up her wounds and wondered aloud how lucky she was.

About 10 hours after the anesthesiologist left the room with a new story, Kristen and I felt as lucky as we ever have. After several hours of athletic effort, she pushed into the world a nine-pound, two-ounce little girl with a full head of dark hair.

SUMMER OF GRAVEL AND STEEL

Anna Kay Rozell has her Mom's long fingers, her nose, and curvy toes that will someday press deep into the tundra. I hope she was also fortunate enough to inherit her mother's sunny nature and resilience. Little AK will someday learn about her Mom's unique qualities. Maybe it'll be when she's nine or 10, and she notices those scars on her mommy's back and shoulders, and she hears the bear story for the first time.

For the record, I didn't think our 10-year-old had asked about those scars yet. But she had listened when Kristen told the bear story to others.

* * *

Bruised and sullen storm clouds spat rain at us. Kristen carried the dripping gun without complaint as I walked next to Anna. As we scuffed along the gravel of the pipeline pad, Anna addressed a subject we had been talking about the last few days.

"Dad," she finally said, glancing over at me. "I know you want me to walk with you from here. But I just want to go back to Fairbanks and enjoy my summer. It would be fun out here with you, but I just can't do it."

Did she hear my heart thud to the ground?

"It's OK, honey," I said, having anticipated her desire to leave. "I think the weather might be getting worse anyways."

Though I was quite disappointed, my weather observation was not just to pacify my daughter. The rain kept falling harder, and ahead loomed a climb over Atigun Pass, where the pipeline crosses the Brooks Range. That led to the North Slope, which has no trees, more bugs, and an uneasy feeling of being constantly exposed.

The girls' week-plus was coming to an end right during an unpleasant stretch of summer weather. The wind whipped around us, driving the rain through our coats and rain pants. Unlike Kristen and Heather, Anna did not suffer in silence.

She screamed to the sky.

"I hate this and I want to go home!"

She shouted what we three adults felt but did not vocalize.

As an added stressor, the tour-bus driver scheduled to pick up the girls that day was missing in action. According to the last report from the satellite texter, he had stopped at Galbraith Lake 100 miles north. His dot on the map hadn't moved southward in several hours. Michelle, in Fairbanks, didn't know why and couldn't reach him.

At one point, the pipe dove under the river and we had

no way to cross the water. Instead, we hiked along the Dalton Highway, its saturated mud as greasy as gear oil. I kept pulling my texter out of my pocket, smearing the raindrops off the screen to read it. No updates on the van.

The rain kept falling. Our body temperatures kept dropping, making us all think of a warm fire that was not going to happen — not only because we weren't stopping to collect firewood, but because those twigs were, like us, drenched.

Anna started sobbing, her face cherry red.

Kristen and I exchanged glances. That's when it struck me: I had seen our girl several times like this on trips. There was perhaps an extra reason she was going nuts: She had to use the facilities.

"Anna come into the woods with me," I said. "You'll feel better."

She did not argue. I had struck the right note.

She followed me into a patch of willows and 20-foot balsam poplars along a creek. They were the last stems of what a person might call forest as the pipeline soon began its climb to Chandalar Shelf.

The Shelf featured one of those drastic landscape changes you feel as soon as you reach it: it's treeless views toward the wall of Brooks Range peaks lets you know you are in the Arctic, far north of the circle we had walked past weeks ago.

Amid the willows and poplars, I dug a quick hole with the heel of my boot. My girl used it. I kicked duff over it and sealed it by setting down a spherical blue rock retrieved from near the highway.

Anna became a different girl. Still not happy about the weather, but no longer as vocal. Her face morphed from red to tan, tears drying on her cheeks.

* * *

While waiting for us in the mud by the highway, Kristen's whole body had started jerking with a shiver. Heather, trying to keep warm, had hiked up and down the road to the North Slope Borough sign: *The World's Largest Municipality*. She was on her second lap. As Heather returned in her blue poncho, with a determined stride that reminded me of our grandmother's, she shook, too.

"I'm going to pitch the tent," I said. "These trees are the last place we can hunker and I'm not sure if that driver is going to show. We can camp here. It doesn't make sense to go up on the shelf. There's no cover there."

I took the girls' silence as grim acceptance.

"We'll stay here on the road and try to flag down a ride back to Wiseman," Kristen said, her teeth chattering.

I retreated to the blue rock — placed on the only flat tent site available. Right next to it, I pitched the tent, still a sodden wad from the rainy day before. The fly felt like a saturated washcloth, but when tightened its wet membrane still shed the rain. We were in for a soggy night — with all three of us jammed into one tent for simplicity — but our moist sleeping bags would prevent hypothermia.

Just as I finished pitching the tent, I felt the low-frequency rumble of an engine, then heard pops of gravel. A vehicle was slowing down on the highway.

I turned to see Kristen in the middle of the road, speaking to the driver of a pickup truck with an extended cab. The truck was pointed in the right direction — south. The driver's nods seemed to indicate sympathy.

Soon, the girls hoisted their wet backpacks into pickup's bed, which held the metal stems of signposts. Pat from Wiseman had scavenged them from somewhere up north. Maybe he used them as fenceposts.

About the same age as me, Pat had friendly eyes and a long beard with patches of gray. He was married to the sister of a man I had met in Wiseman. I had somehow never met Pat in that small town.

My shotgun also went into the bed of Pat's truck as the girls prepared to leave me. Though I felt comfortable hiking without the gun, the sense of security that came with it had gotten inside me. I felt a pang at the gun heading southward.

Mainly, though, I felt the loss of the companionship of the girls. But the time was right; a cold, wet night lay ahead for Cora and me. The girls had endured enough.

I exchanged wet hugs with Anna and Heather. I kissed the top of Anna's head. They then crammed into the back seats of the truck. Pat cranked the heater fan to high. "Turbo Lover" by Judas Priest played on his speakers. I laughed out loud.

"I know that song," I told him.

I have seen Judas Priest a few times, most recently long ago with my brother Drew in Troy, New York. Lead singer Rob Halford rode his Harley on stage, destroying our eardrums by revving the motorcycle. Then he pulled out a bullwhip and started flogging the machine.

I felt like I had known Wiseman Pat for a long time. The girls were in good hands.

After giving me a kiss that tasted of rainwater, Kristen climbed into the passenger seat and slammed the door shut, sealing out the weather. Pat pulled away. Heather turned and waved through the rain-streaked back window. Anna never looked back, which I noticed. Had I pushed her too far, again?

The hiss of the tires in the mud faded as the truck got smaller and smaller. The narrow valley just south of the Continental Divide grew quieter.

My only companions now were Cora and the droplets that kept falling. Cora followed me toward the rain-wrinkled tent. It promised the moist warmth of a sleeping bag. I kicked myself into the bag. Cora backed herself against me on her pad. Soon, I couldn't feel the dampness anymore. I relaxed and fell asleep to Cora's breaths and the patter of rain on the fly.

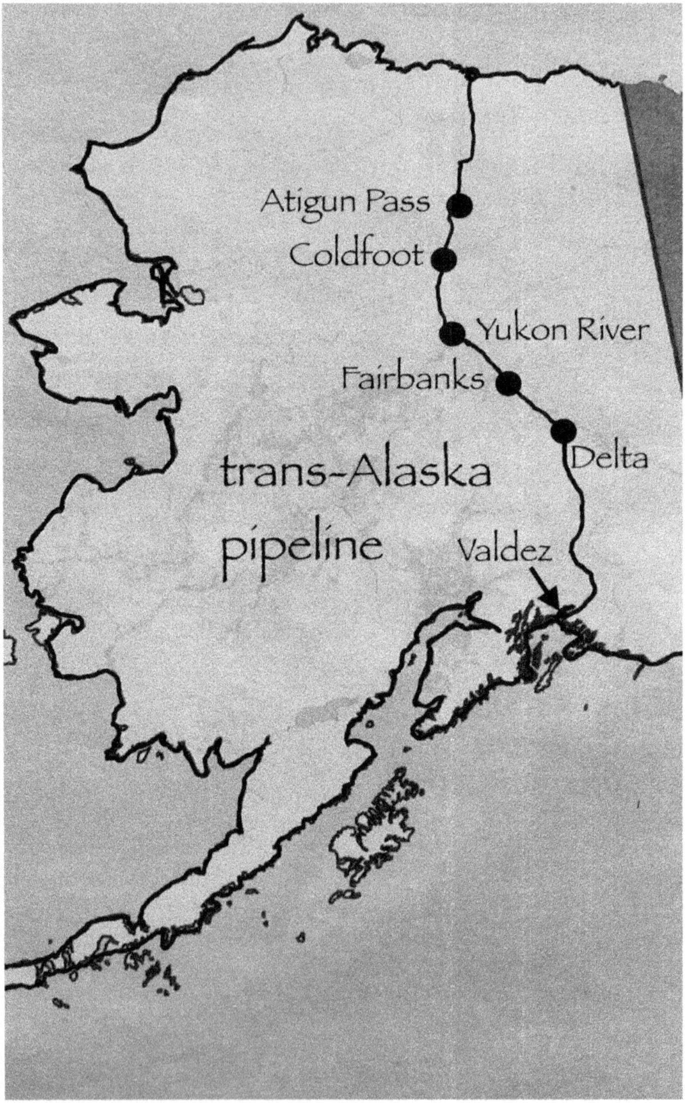

Atigun Pass
Coldfoot
Yukon River
Fairbanks
Delta
trans-Alaska pipeline
Valdez

21

Exit the boreal

This is a story I wrote while sitting on a rock after crossing the Continental Divide, the day after saying goodbye to the girls. I wrote more than a dozen of these during the trip and sent them back to my colleagues at the Geophysical Institute, who sent them to newspapers and websites.

ATIGUN RIVER — Goodbye, red squirrels.

On our summer-long hike along the path of the Trans-Alaska Pipeline, this morning my dog Cora and I left the last tangle of boreal forest along America's highway system. We walked away from a campsite of white spruce and balsam poplar that shielded us during a rain and wind storm the day before.

The squeak we heard from a red squirrel, whose diet is

mostly spruce seeds (but occasionally fledgling birds and baby snowshoe hares), was the last we'll hear until we return home to Fairbanks when this adventure is complete.

Following the Dalton Highway and heading north, we walked up a few thousand feet to Chandalar Shelf. Willow shrubs and alder, yes. But the large trees were no more.

It took a long time to out-walk the boreal forest.

Since we first saw aspen trees along our route just south of Copper Center, Cora and I have been moving for two months to transit that band of oxygen-producing, carbon-storing scratchy plants. On this continent, the boreal forest extends from western Alaska all the way east to the Maritime Provinces of Canada.

As we followed the pipeline's path through Atigun Pass and crossed the Continental Divide at about 4,500 feet, we stepped into a new world.

Here on the north side of the Brooks Range, the misty mountains spill clear-running creeks. From where I sit with my back against an industrial metal shed related to a pipeline valve, I hear the worried shriek of a peregrine falcon. It is a greeting to a land with yearly temperatures cold enough to prevent the invasion of trees, a place where winter is the norm and summer visits for just a few months, then scurries away.

I live in the boreal forest and am comfortable in the poplars, aspen and spruce and on the rivers that twist through them. Crossing the pass, which is the highest point on the pipeline's route but far from the steepest climb or descent, I entered a rainy, cloudy, treeless world. Mystical is a word that keeps coming to mind. It is the same sensation I remember from 20 years ago, when I crossed over with my brother-in-law James Hopkins.

On the other side, in that Iowa-sized part of central Alaska known as the Interior, I leave the 80-degree days we experienced from the Yukon River all the way to the base of Atigun Pass. Goodbye moose flies, dunks in clear rivers, hot nights in the tent and, at the bridge, tanned kids and their parents arriving by riverboat from fish camps on the Yukon River.

And goodbye thunderstorms. A few days ago, near the site of the old Dietrich pipeline construction camp, lightning struck so close to my wife Kristen, daughter Anna, cousin Heather Liston and me that we heard the crackle of static and instantaneous thunder. We took cover under two U-shaped forms of concrete sometimes used to weight the pipeline as it goes under rivers. Lightning happens here on the North Slope side of the mountains, too, but it's rare compared with the heat-driven convection cells of the Interior.

Heather and Anna take shelter from lightning

What to expect here, where the Atigun River flows northward, joins the Sagavanirktok and heads through the bumpy flats of the North Slope to gray saltwater? Stunning mountains, for a bit. Cold feet, because I'm still wearing wet running shoes. Caribou chewing lichen. Wind. A visit to Toolik Field Station, where my neighbor and UAF grad student Jason Clark will warm the sauna.

And of course, more mosquitoes. Though the wetlands of the Interior were impressive when traversed with improper timing, the North Slope mosquito is the queen of them all. She is the type that inspires competition. How many can you kill with one slap?

Out of respect for her, I did not rush to the north side of the divide. With my family and cousin from San Francisco, I walked slowly through the spear-like spruce, sculpted white mountains and aquamarine water of the Dietrich/Koyukuk river country. I wanted them to see what I considered the nicest part of the trek 20 years ago (Coldfoot to Atigun Pass). And I wanted the North Slope mosquito to be on the waning end of its few-week life cycle before I dropped in. I hope I'm late for the party.

22

Into the Anthropocene

We arrived just in time for the party. A slight tailwind kept my friend Eric Troyer and me smack in the middle of our own mosquito entourage. Our exhalations invited more and more bodies, as did Cora's recruiting trips to the brush, her back furred with mosquito wings.

While annoyed, we knew the situation would soon change, when the wind shifted or the pipe's pathway turned. Until either happened, we would continue absorbing pinpricks.

Just a typical sunny August day in the far north. A good time to practice the philosophical wisdom of impermanence.

Mosquitoes live only a few weeks; their insane summer riot was near its end. Then it would be hard to remember they even existed.

Most mosquitoes die soon after dabbing eggs in a wet spot. The females, that is. The only ones who feed on blood. The only ones most of us care about.

But not all female mosquitoes die after laying eggs. A few flutter to the tundra, where they overwinter beneath a blanket of snow. The first insulating flakes were less than a month away.

But the riot's end was still in the future. Eric and I were living in the present, trying to accept the inevitability of our situation with patience. We did not always succeed. Some bad words rung in the summer air when we felt the stings that had slipped past our chemical and physical defenses.

But we did not always feel it when those little proboscises penetrated our skin. Over the years, we have both spawned many generations of mosquitoes with our blood. Eric a few more crimson droplets than me.

A lifelong Alaskan and son of pioneer biologist Will Troyer, Eric grew up accompanying his father on many personal and work trips. Those included moose hunts, sheep-counting missions, and a broad survey of quiet places for possible designation under the Wilderness Act of 1964.

Eric and Cora

Moving with Will's work assignments within the U.S. Fish and Wildlife Service, the Troyer family lived on Kodiak Island and the Kenai Peninsula before getting sucked in by the Alaska metropolis of Anchorage. In all three places, the family spent many hours outside the house.

SUMMER OF GRAVEL AND STEEL

Eric's pedigree showed on the trail with me — he knew how to pack. I could pick up his backpack with my index finger without straining a ligament. I wondered why my pack felt so heavy. I had hiked three months with the load pressing down on my shoulders. Had I really needed all that stuff?

We walked on, waving away mosquitoes and scuffing our feet along meatball-size rocks, some embedded with flakes of blue and red. All had been dug from a particular far-north quarry and then dumped here on the tundra to provide a path for pipeline construction and maintenance workers.

With every step we took into the lumpy green lowlands ahead, the purplish mountains of the Brooks Range grew ever smaller behind us. Cora and I would walk through no more mountain ranges during this hike. Now, from the higher part of the North Slope, the path ahead lay toward the increasingly flat coastal plain and, finally, the industrial terminus of the trans-Alaska pipeline.

Eric had driven about 400 miles north from his home in Fairbanks through Atigun Pass to join me. Like many Fairbanks people, he views the Dalton Highway with respect, but not as the death-defying drive sometimes portrayed on reality television. He made the trip in his two-wheel-drive Prius, an extra jug of gas strapped to the roof and a full-size spare tire in the back.

A few years older than me, Eric is cheery without fail. His voice booms, often punctuated by a laugh. Sometimes he is too well modulated for my tin left ear. I damaged it many years ago by firing a modified .338 magnum rifle at a spring caribou on a hillside near where we were hiking. I missed the shot because my scope was misaligned. And because I flinched.

A gunsmith had "ported" the barrel of the rifle with precision-drilled holes to reduce the kick; his work, performed at my request, had the side effect of making the gun twice as loud. I hated pulling that trigger, so much that I sold the gun after that hunt.

My left ear now rings almost constantly. Some frequencies and volumes hurt a bit. This caused me to ferry across the pad to keep Eric to my right.

I met Eric a few decades ago when he was features editor for the Fairbanks Daily News-Miner and I was a cub reporter, still in journalism school at the university. I liked his supportive, friendly style of critique so much that I chose him to edit this book.

Eric did not continue as a professional journalist. Not because he could see the digital future and the ultimate demise of the paperboy's way of life, but because he found other goals. He decided to teach elementary school and got his teaching credentials. Along the way he met his wife, Corrine

Leistikow, a doctor (my doctor!) After they had two children together, Eric made the call to become a househusband.

But the writing bug has not left him. Eric has freelanced, and is volunteer editor of two trails newsletters — one Alaskawide and another focused on the Interior where we live. He advocates for dirt and sintered-snow pathways everyone can use, attends overlong local meetings, and has improved the quality of life for all people in our town, whether they realize it or not.

With both his children fledged and living outside Alaska, Eric has more time for the outdoor adventures of which he and Corrine are so fond. He said yes when I asked him to share a few miles of pipe.

* * *

As we hiked, to our right loomed an immense block of Alaska of which almost every American has heard — the Arctic National Wildlife Refuge. This swath of land includes forests, a large chunk of the Brooks Range, and part of the North Slope's coastal plain that includes the controversial "section 1002."

I don't think about ANWR much because it lurks so quietly amid so many thousands of other silent acres. But the land has become a symbol of a battle that comes to mind when I'm driving next to other cars in Yonkers. It's a fight between preservation of unpeopled land and our need for a volatile liquid that powers our cars, heats our homes, and enables us to fly to Chicago for a Bruce Springsteen concert on a whim.

For as long as I have lived in Alaska, people near and far away have argued about the quiet hills to our right, and the plains north of them that flatten as they stretch toward the Arctic Ocean. The main argument centers around whether to open it to oil exploration.

Natives and white explorers have been finding oil on the North Slope for years in the form of seeps and tar sands. In the 1940s, members of the American military started hunting for the black liquid, but it wasn't until 1968 that oil-company drillers discovered vast reserves near Prudhoe Bay, which spurred construction of the trans-Alaska pipeline.

In ANWR's coastal plain, only one test well has been sunk, back in 1986. Results of that test proved "worthless," according to a reporter who wrote a story in the New York Times in 2019. But drilling advocates claim the area still has great promise.

I once interviewed biologist and conservationist George Schaller, who visited the mountains and coastal plain of ANWR on a trip in the 1960s, when it was called a range instead of a refuge. Schaller had attended college in Fairbanks back then; people knew him as the guy walking around campus with a pet raven on his shoulder.

When I met him in 2006, Schaller had just repeated the same journey to ANWR.

About that trip, taken when he was 73 years old, Schaller told Smithsonian writer Sasha Ingber:

"I wrote Secretary of the Interior Fred Seaton in 1957 and said, 'Hey, you've got to protect that area.' And thinking about the oil exploration camp I had already seen on the Arctic Slope, I said, 'That area may well in future years resemble one of the former Texas oil fields.'

"I couldn't believe the horrendous environmental damage there when I visited in 2006. It's 800 square miles of buildings and roads and pipelines and drilling pads and oil spills. It's a dreadful place. It will never, ever be fixed up. So now is the time to protect the coastal plain. It has been a horrendous battle since Secretary Seaton established the Arctic Wildlife Range [now the Arctic National Wildlife Refuge] in 1960."

As Eric and I hiked, the horrendous battle continued in far-away halls and conference rooms. There, politicians and activists speechified, rallied and chanted. Near the boundary of the refuge, along the pipeline, Eric and I heard only the whine of mosquitoes and the rumble of far-off trucks on the Dalton Highway.

For now, the environmentalists have won: No buildings, roads, or pipelines exist on the coastal plain of ANWR. Caribou cows still drop their spring calves in the same breezy, wild place their ancestors did thousands of years ago.

Eric is one of the relative few to have traveled in the refuge and coastal plain. When he was 12, he spent a couple weeks there with Will, who was surveying the area for possible designation as Wilderness. Eric remembers a migrating caribou herd flowing around he and his dad. The animals were mostly quiet, except the finger-snap click of tendons in their feet as they walked.

Though we didn't talk about ANWR, I would guess Eric feels the same as I do: We know the conveniences of our modern life are shoving billions of organisms from their preferred habitats. It causes a twinge of guilt. Leaving a little space seems the least we can do. Schaller felt that way.

"Animals and plants have a right to exist, too," Schaller said to me when I got to interview him at the university in 2006. "We don't have to destroy them for short-term goals. Ultimately, our quality of life depends on a healthy environment, and for a healthy environment we need open space."

After we waded a thigh-deep creek at pipeline mile 100, Eric and I saw some caribou in the distance, dipping their regal heads amid a sea of green tundra plants. To the east their kindred in ANWR traveled in open space, so far free of pipelines and gravel roads.

* * *

As we hiked, the sky drizzling on us, Eric, Cora, and I were truly on the North Slope, the final 80 miles of the pipeline hike running over flatness that led to the top of Alaska and then the Arctic Ocean.

After a few hours, we strode into a new age: The Anthropocene. Or at least its consequences.

The land on either side of the pipeline's gravel road had sunk in an odd way. An uneven chasm the size of a village school dipped 20 feet beneath the surrounding tundra.

As Eric walked around the edge of the sinkhole, I snapped some photos of him for perspective. In the vastness of the North Slope, things can seem either huge or tiny without something nearby for comparison. (For example, I once mistook a ground squirrel for a far-off grizzly — until the "bear" squeaked and disappeared into its burrow.)

Someone had pounded wooden stakes into the ground surrounding the large depression that nearly devoured the road and, just beyond it, the pipeline. A nylon measuring tape wound through the stakes.

Later, a full-size pickup crept up the pipeline pad toward Eric and me. The rain-streaked window lowered to reveal the dimpled smile of Margaret Darrow. Mikhail "Misha" Kanevskiy sat in the passenger seat. I knew them both from their work as permafrost scientists, Margaret working with the state and Misha the university.

Misha and Margaret

Margaret called it a "retrogressive thaw slump."

Many of these giant sinks in the northern landscape had appeared the past few years, keeping scientists like Margaret and Misha busy. The depressions open when mammoth-size chunks of ice melt far beneath the ground surface. As they melt, they create space into which the ground above sinks.

These craters developing in northern lands indicate a new era some call the Anthropocene, an epoch named for the billions of environment-altering creatures who resemble me and you.

Thousands of years ago in a colder time — before we grew so numerous and fossil-fuel dependent — sustained bitter air temperatures penetrated deep into the ground. The cold transformed groundwater into ice, or more often, an ice-and-soil mixture. This "permafrost" — ground that has remained frozen through the heat of at least two summers — formed in huge areas across the far north and far south (though it's less common on the bottom of the globe because there's less land).

Even when global temperatures warmed after the last ice age, the permafrost stayed frozen under a blanket of insulating tundra, mosses, and other plants.

We humans have used our evolving brains to defeat this ice when we want to go beneath the surface, such as while mining gold or digging a traditional Native ice cellar (a "siġluaq"). Heat is a common tool. We peel back the blanket and let in the sun, or build fires to melt the ice in the ground beneath. I've done the latter when digging an outhouse hole.

But now these thaw slumps appear even in areas where we have not stripped off the natural blanket with our shovels and dozer blades. Air temperatures north of the Arctic Circle have risen, on average, about four times as fast as temperatures in Texas during the last few decades. The warmer air temperatures penetrate the ground and thaw the permafrost.

Massive divots now pock the face of northern Alaska, Canada, and Siberia. And they don't cause problems just for metal pipelines. One slump, on the bank of the Selawik River in northwestern Alaska, clouds the waterway with the sediment it has freed, making it less desirable for whitefish that prefer to lay eggs amid clear cobbles. This could reduce survival of the fish, an important food source for the villagers of Selawik.

While that particular sinkhole may now affect only a few hundred people in the far north who harvest a certain fatty fish, the craters as a whole cause impacts that touch people across the world, as far away as Bangladesh, South Africa, and Brazil.

As Eric and I explored the slump near the pipeline, we inhaled emissions that are probably the biggest impact of these features: methane. The gas was too diffuse to poison us, but it is extremely powerful when added to the gases changing our planet with more than double the heat-trapping powers of carbon dioxide.

Where does the methane come from? When permafrost thaws, organisms that have slumbered for a long time wake up. These stimulated soil microbes are hungry as anything else.

They feed on suddenly available plant and animal bits that also had been locked in ice for a few years or a millennium.

Just like us, the microbes offgas. Also, just like us, a large percentage of their emissions are methane, which mixes into the 30-mile shell of gases surrounding the planet.

While checking out the slump, I couldn't help but to think of the irony: Ground thawing for the first time in thousands of years now threatened the stability of the pipeline, a tube that has transported so much of that climate-changing liquid during the past half century.

That Eric and I had walked right into this new emissions hotspot was not a total surprise when we thought about it for a while. On the skullcap of the planet are hundreds, maybe thousands of these new sinkholes, some much bigger than this one.

The pitted face of the far north is daunting to think about in an existential sense. Eric and I and everyone for whom we care live in the middle of this drastic change in Earth's climate that has been somewhat stable for many generations.

This change seems to be of a similar magnitude as what happened after the Last Glacial Maximum, a time when the great ice sheets melted, sea level rose, and species including the woolly mammoth went extinct by the dozens.

Scientists have predicted today's extreme warming, but much of the effects — like permafrost thaw — happen in slow motion, kind of like the gray hairs sprouting on your head; you wake one day and notice they are the majority.

True, the planet has experienced drastic temperature changes for as long as it has existed. Millions of years ago, where Eric and I hiked, winter was just as dark but not nearly as cold. A lush world of green ferns waved beneath water-loving trees like those now growing on coast of Oregon.

We have lived in a sweet spot for humanity for an overlong time.

"Since the ice ages ended 12,700 years ago, Earth has existed in an unusual state of climactic calm," wrote Dan Mann, who studies ancient landscapes in Alaska and has taken me on a few trips. "Instead of experiencing climate flips between full-glacial and near-interglacial conditions every few millennia, we've had a long period of relative stability. This anomalous period of relatively stable climate has facilitated the rise of agriculture, the explosion of the human population, and the consequent impacts of these developments on the biosphere."

We now live on the edge of a new, much warmer world than our species has ever experienced, and we have helped speed our way toward it. And, as Dan points out, our ancestral habit of not looking beyond our next meal makes it both easy and natural for us to kick the can down the road.

We have released into the atmosphere so much greenhouse gas that the climate will get warmer for at least decades to come. Glaciers will melt, tropical storms will intensify, weird weather events will happen more frequently, low parts of coastal cities will be ankle-deep in salt water again and again. But some creatures, such as certain species of mosquito, will like the changes.

Like every scientist I have ever asked, I'm not sure what to do about runaway climate change. The problem is too big. I feel the same paralysis when I contemplate replacing the foundation of my home, which sinks a few inches farther from Heaven each year due to permafrost thaw. So far, I have chosen the Homer Simpson Solution: Hide under a pile of coats and hope that everything turns out all right.

Even if I don't know what to do about the future of humanity — besides thinking the world would be better off without so many of us — I am an optimist. Humans are clever and adaptive. And the future never really unfolds as we predict, does it?

23

Classic rock

There in the distance, we saw it: A blond lump rolling over the tundra. Grizzly bear.

We watched as it traversed a bushy hillside a couple hundred yards away before disappearing into the greenery.

"That's the first one I've seen on the trip," I said to Eric.

"Better late than never!" he said.

I appreciated having unfazed Eric along. The bear wasn't just the first grizzly I had seen on the trip. It was the first bear of any kind. Seeing it made me nervous.

We were planning to camp on an open gravel pad, not far from where the bear had vanished into the bushes.

Maybe Eric's childhood growing up with a bear biologist makes him so non-afraid of bears. I found myself scanning the nearby horizon with each lift of my head as we pitched our tents. Eric continued his breezy banter.

Still, even Eric seemed more wary than normal. Earlier in the day, while we talked to Margaret and Misha about the thaw slump, they warned us of a limping bear in the area. Now Eric and I wondered if it was the same bear we had just seen. We hadn't detected a limp, but we hadn't gotten a good look. The bear had been moving fast, so maybe — hopefully — it was a different bear.

We ate dinner sitting on rocks with our faces turned to a blessed light breeze, keeping the mosquitoes behind us. I chewed lasagna with my head swiveling. In my journey over seven-eighths of Alaska, I had come to believe that bears no longer existed. Their lack of presence made them seem ethereal. Not anymore. That image of the rippling blond coat wouldn't leave my head.

***

That I had hiked across 700 miles of Alaska — almost always far away from cities and towns — and hadn't seen any bears seems impressive. But that's the way it is with bears, especially in the Interior and Arctic where they roam impressive distances and there's a lot of distance to roam. You might see three in a weekend or you might see none for 700 miles.

The ones you do see almost always avoid you. I knew they were around. Plenty of tracks and blueberry-filled scat, but I hadn't seen one until now. That's why I carried the shotgun only when I traveled with Kristen, who has her excellent reason for wanting one along. I wouldn't have minded carrying it after seeing that grizzly.

After many miles of hiking, Alaska was still surprising me. And it wasn't finished.

* * *

Our food-storage method for that night was lame. For many nights on my trek, I used my companion, the pipeline, for food storage. I would toss a line over the pipe, haul up my food bags until they lay against the pipe near the top, then tie off the line to one of the pipe struts. The food hung safely out of reach of bears or other foraging animals.

At campsites where the pipe ran underground, trees almost always grew nearby. I would throw a line over a high branch or climb up to store my food. During all those nights, I had no problems. My food remained as I had left it.

There in the north-of-the-Brooks-Range land of permafrost, the pipe typically rode high above the ground. But in this spot the pipeline traveled underground, buried in the unfrozen gravel of the Sagavanirktok River. I had left behind the boreal forest long ago. No trees. Food storage options were less than optimum. Which wouldn't have bothered me much if we hadn't just seen a bear.

Fortunately, near our camp, a sign on a metal pole stuck out of the ground like a seven-foot T. The horizontal part displayed the number 76, the mileage from us to Prudhoe Bay.

I found a pallet nearby then leaned it against the sign as a makeshift ladder. I stood on top of the pallet, tied the food bags to the sign, climbed down, and dragged the pallet away.

The food bags hung suspended, their bottoms only six feet off the ground, away from fox and other small foragers, but within the reach of any adult bear. Not secure, but better than in our tents or sitting on the ground. Eric eyed the food bags and shrugged. It was the best we could do.

Later that night, within the tent, Cora growled. I woke.

She sat up and sniffed at the fresh air vent above my head. The hair on her back stood straight up. Adrenaline button pushed.

Cora exploded with barks. Something was out there. Something more than a vole or a hare, but I couldn't see through the tent walls. I thought bear. The same bear? The possibly hurt bear?

I sat up. Dressed only in my underwear, I unzipped the tent door and fly. Cool air flowed in like water while Cora shot out like a cannonball, barking all the way.

"What's up?" Eric asked from his nearby tent.

"I don't know."

I pulled on my pants, feeling stupid for letting Cora out. What if it was a bear? Had I sent her right into its jaws? Cora's barking stopped in the distance by the time I stood up on my sneakers, staged outside the entrance. Her quiet concerned me, but at least she wasn't shrieking.

Pepper spray in hand, I stepped into my shoes. Thinking of our half-ass food storage method, I walked softly toward the sign, hoping that it still held our food. Eric followed, carrying his own pepper spray. Mosquitoes bumped off us like air molecules.

At the end of summer, arctic nights are still somewhat illuminated, but the midnight sun no longer stays above the mountains. We could see without headlamps, but the twilight felt thick, the dimness adding to the tension. I scanned the area, searching for Cora and any large dark shapes. Even chatty Eric remained quiet.

When we reached the sign, we saw Cora, unharmed and unfazed, sniffing at its base. The lumps that were food bags hung, undisturbed. I relaxed a bit, but we still didn't know the cause of Cora's explosion.

Eric and I squinted at the ground. There, on the compacted soil, we saw five chisel marks in a tight arc. The track had not been there when we had hoisted the bags. Those tracks led away, spaced far apart. A grizzly must have investigated our food and then took off sprinting when it heard Cora bark.

I relaxed a bit more. Running from a barking dog is good grizzly behavior, at least from a human perspective. I clicked twice with my tongue to Cora. She trotted over. I kissed her forehead.

We retreated to the tents, mosquitoes riding our draft. Cora and I crawled inside ours and I zipped up the screen. We lay down and curled into our usual positions. It took a while for my adrenaline level to drop, but my breaths gradually deepened. Cora was out before me. We slept on the ground for the ninetieth time.

* * *

Bears weren't the only mammals I had seen few of on this trip. Earlier, when Eric, Cora and I approached Pump Station 3, one of a dozen run by the Alyeska Pipeline Service Company, I had braced myself for the busyness of human contact.

There, 20 years earlier, workers at the oil-pushing facility welcomed me and Jane with a fuss. The greeters included my college friend Chris Bias — working there two weeks on, one off — and a cook who handed me two styrofoam containers of fried chicken and mashed potatoes.

As the cook handed me the food and I thanked him, a woman with a clipboard stood nearby on the gravel watching. She soon explained that the Alyeska brass had charged her with making sure I was doing OK when I reached the chain-link fence surrounding the pump station. Jane and I were her assignment that day. I hoped she didn't notice my wide eyes of overwhelm as I thanked her. We were her assignment? For the day? I was in a bushy state of mind, feeling more comfortable with birds and ground squirrels than with humans. That was common when Jane was my only social contact for long periods.

My first trip had many interactions like that. Jane and I would travel for days and miles with just ourselves or one or two other people for companionship — almost always friends and family. The introvert in me felt comfortable in that setting.

But as the first documented person to hike the trans-Alaska pipeline — and me writing about it in a column for statewide distribution — I had a certain status. People recognized Jane and me along the way, especially workers associated with the pipeline. Jane was the real celebrity. We would approach, and they would call out her name. But then they would talk to me. Sometimes several all at once. At least, that's what it felt like. And Alyeska executives had bestowed VIP status on us. People within the bustling pump stations were expecting us, looking forward to our arrival. The abrupt change from hermit-like existence to celebrity jarred me every time.

Twenty years later and three months into my current hike, I braced for a similar reception at Pump Station 3. As we approached, we could hear the roar of the turbine engines, suggesting that the station still propelled oil toward the Brooks Range (to be pushed over the top by Pump Station 4). That roar seemed to presage the busyness to come.

To my surprise, the guard shack out front sat vacant. No white trucks rolled around behind the fence. Eric and I looked around, scanning for signs of human life, but saw none. That absence left me unsure of what to feel. Relieved, yes, but also a little disappointed.

I should not have been surprised. Reduced oil flow in the pipe meant a reduced workforce. Hadn't I learned that way back at Pump 12?

I described to Eric the fanfare that Jane and I enjoyed at the same spot in 1997. He shot back a metaphor.

"Back then, you were selling out stadiums," he said. "Now you're classic rock."

I cracked up. "Kashmir" played in my head for the rest of the day.

* * *

Bears and humans weren't the only things keeping my trek interesting. An outside observer might think, as I neared the end of my trek, that I would have few if any gear surprises. I had been on the walk 90 days and nights, plenty of time to figure out what I needed, what I could do without, what worked, what didn't.

Some things I did have figured out. I could set up and take down my tent without much thought. I could find a spot to pitch the tent within a minute of searching. It took not much longer to sling parachute cord over the pipe and hoist up my food. Still, every new companion who joined me traveled outdoors a little differently. They often gave me ideas on how things could be done a new way or gear might work better.

And sometimes gear surprises came from bad luck.

The night after we passed the ghost town of Pump Station 3, Eric and I camped on the gravel 100 yards from yet another pump station. This one, Pump Station 2, seemed even quieter than Pump 3. A family of ravens nesting in a microwave tower were the only moving creatures. The lack of people in the far north was becoming a theme; was it happening anywhere else on Earth?

Eric at Pump Station 2

While sitting on my sleeping pad, half out of my sleeping bag, I leaned forward and unzipped the tent door. Cora, eager to get out but not unusually excited, watched me. Just as I finished unzipping the door, she thought she heard something outside. Cora blasted out of the tent.

BLAM!

A shotgun blast? But in that same instant, my butt felt the hard gravel under the tent floor.

I looked down to see a rip in the orange fabric of the ultralight air mattress Kristen had loaned me. A four-inch gash. Right next to a seam.

I had coveted that mattress because it weighed next to

nothing and took up little pack space. It rolled into a bundle as small as a Nalgene bottle. It was a bit noisy, crunching like potato chips when I rolled over, but Cora didn't seem to mind. I borrowed it when I stopped in Fairbanks. I wanted to save space and lighten my load. Kristen could have used it that summer, but she loaned it to me even though it was her favorite.

I made a couple of attempts to patch it. Eric watched me work and then offered to try when I gave up. He valiantly used all his fabric patches, but the gash was too long, the placement of the rip too awkward. The mattress had become dead weight.

That night, I pitched the tent over a soft patch of moss, sleeping without a pad for the first time on the trip. I hadn't looked forward to the night, but the moss worked well enough. Not as cushy as the pad, but I slept fine.

The next morning, Eric and I cooked our breakfast in the chilly air using one of the support struts for the pipeline. It stood at kitchen counter height and provided a flat surface for our stoves and pots.

But my stove wouldn't light. I tried several times without success, but it wouldn't exhale gas when I turned the valve. I had used that same stove for my entire hike without problems. Nothing obvious seemed wrong.

Eric's JetBoil stove worked fine. I already coveted that stove, which has an integrated pot and windscreen. It boiled water quicker than my cook kit.

On Eric's last day, we hiked to an access road and out to the Dalton Highway. There we waited for Kathy, a driver with the Northern Alaska Tour Company. She would give Eric a satellite-text-arranged ride back to his car near Toolik Field Station.

Before Eric departed, leaving me to be a solo artist for the final time of the trip, he loaned me his air mattress, JetBoil stove, and lightweight sandals for crossing the waterways I anticipated over the last few days. Did I now have a system that was perfect? Probably not, but I was getting more dialed in. I again appreciated learning something from a hiking companion.

After Kathy pulled over in a white van, filthy with Dalton Highway dirt, Eric and I hugged and said our goodbyes. I would finish my hike with Cora as my sole companion. I had been happy to share trail time with Eric, but — as happened every time — I was excited both when my companions arrived and when they left me.

After a wave to the departing van, Cora and I headed back to the pipe and its gravel pad. Man and dog had a little more of Alaska to experience, its grandeur, its serenity, and probably another surprise or two.

trans-Alaska pipeline

- Prudhoe Bay
- Atigun Pass
- Coldfoot
- Yukon River
- Fairbanks
- Delta
- Valdez

24

The end

Mile 10, trans-Alaska pipeline, August 8

Magical, quiet night on the true coastal plain.

I'm camped exactly where I camped 20 years ago — behind the pipeline anchor point closest to Mile 10, which indicates we are 10 miles from Prudhoe Bay and the start of the pipeline.

This spread of roundish rocks not much bigger than the tent's footprint is the only place to pitch. The gravel dumped by front-end loader in the 1970s is the only place that's not lumps of tundra plants with cold black water in the holes between. There simply is no other flat space here on top of the world.

It's a fitting place to end.

The weather is much better than 1997's August blizzard that blasted snow into my eyeballs. The view is expansive of this wide-open country. From 10 miles away, the lights of Deadhorse look like Las Vegas as viewed from the surrounding desert.

Devoid of man noises out here, where the pipeline has flared away from the highway. White-fronted geese are laughing as they walk their orange feet on the gravel of the pad 100 yards away. Cora is aware of them but is looking at me for dinner.

There is miraculously no breeze and no bugs. Unreal up here in the farthest north of Alaska's Great Swamps.

When I write tomorrow night, everything will have changed. Cora and I will probably be sleeping in a hot room on the grounds of the oilfields. I will need a shave.

Since snow was on the ground, Cora and I have moved, using our little feet, all the way across Alaska.

To make this final trail night even more memorable, I saw a grizzly after I had set up the tent. He was there before, but I didn't notice him until he moved.

He was blond from a summer's worth of sun rays, close

enough that I didn't need to enlarge a photo to see what he was. He was not limping. I wished Eric was still with me.

To alert the bear to our presence (Cora had not detected him), I picked up a palm-size rock and clanged it against the rusted steel of a hollow vertical-support-member pipe.

Bong, bong, bong, bong.

His head shot up as the sound waves intersected his ears. He looked around, not able to detect me or Cora.

I remedied that by walking back toward the previous pipeline support. I wanted to force the interaction, not delay it into a midnight surprise should he wander closer. A very slight breeze was blowing from him to us, nullifying his nose.

The bear noticed my horizontal movement, locking his eyes on me as I walked.

He made a decision: He chose indifference. He did not move closer to explore. Nor did he move farther away. He dipped his head to absorb the smells of vegetation he was exploring and then tore up some roots and chewed them.

I stood watching him meander peacefully for an hour, a golden smudge bobbing amid tussock heads.

Then, he with great speed bounded across the tundra,

heading toward one of the giant oval lakes between here and the Dalton Highway. Looking at my gps map, the lake is three miles away from the tent.

It will probably be hard to sleep tonight. And tomorrow we will be done.

* * *

The barren-ground grizzly did not wander our way in the night. I did not need to throw any of the 10 baseball-size rocks I had staged in a row on a pipeline support beam.

In the morning, after I pulled out my trail ear plugs for the final time, we exited the tent.

Both Cora and I then peed on tundra plants. It was windy but warm enough as I used Eric's JetBoil to make my coffee water in record time. The lake where the bear had been rooting around was again just one of thousands of cigar-shaped lakes on Alaska's North Slope. No golden smudges that I could see.

I stuffed my pack and clicked Cora into hers for the last time. That morning, August 9th, marked our 96th awakening on the ground since stomping through the rotting Valdez snowpack in late April.

Hey, it took me 120 days to cover the same 800 miles in 1997. FKT baby!

Though, as my athlete friends would acknowledge (as would I), this new Fastest Known Time for hiking the trans-Alaska pipeline was a soft record. Any relatively fit outdoor athlete intent on breaking it could crush it easily. Still, I appreciated that I was the sole person known to have done it. Now, twice.

As I walked the gravel ribbon toward the industrial complex that marked the finish of the trip, I remembered the words of Andy Sterns as he and I neared completion of a spring ski trip from Nenana to Nome.

"It's no great shakes getting to Nome," he said.

Andy and I felt more anxious than thrilled as we skied up to a gas station in Nome, marking the end of our ski trip so many years ago. We started the 674-mile trek by skiing away from a gas station — and the road system — in Nenana.

When out on the trail that long, your life experience takes on a different hue. You get used to the sublime simple life of everyday existence. Approaching the end of your journey, you start to remember complex, fast-paced modern life with its conveniences, but also its overabundance of choices and stresses and noises. Hence, the slight anxiety, just as you feel when floating a quiet creek and hear the roar of whitewater ahead.

The unknown lay ahead for Cora and me: I didn't know what to expect at trans-Alaska pipeline mile zero. I had made no arrangements for a resting place with the oil-service company. And the fanfare continued to be as silent as the wind whispering through the tundra grasses and rippling the lakes. I imagined I could smell salt water. Maybe I could.

Twenty years earlier, an Alyeska pickup driven by an official I had befriended carried John Arntz, my San Francisco friend and sometimes hiking partner, in the shotgun seat. They came out to greet me and my girlfriend Clara Jodwalis as we hiked the last few miles in a wet late-August snowstorm. It felt like a big deal to be officially greeted.

But classic-rock pipeline hiker kept walking, wondering where he and his dog might sleep that night in a humming industrial complex.

Then, a pleasant surprise: Little footbridges!

Twenty years earlier, a few times in the last few miles, Clara, Jane, and I had encountered moats of cold water crossing the gravel road. Knowing the day was our last, we waded through each of them in cavalier fashion, drenching our leather boots and soaking our feet in frigid water. Our bodies wicked up the cold then, and a constant breeze stole our heat. But we knew we were almost done.

To my delight, Cora and I found plywood bridges, complete with handrails, over all the water-filled trenches. I later learned that Alyeska had recently staged a 5-kilometer fun run on the pipeline pad. Shop workers had deployed the bridges to keep everyone's shoes dry.

* * *

One last trail surprise: As Cora and I walked past the orange sign with a black "2," a truck pulled over on an adjacent road that was part of the nearby Pump Station 1.

Two people exited the truck wearing rubber boots. They started soft footing the wet tundra toward us. They were too far away for me to make out their faces or any markings on their coats. Security? No, too much of a relaxed pace.

Then, two friendly waves.

The pair got to within about 100 feet when Cora finally saw them and barked.

Most unusual, these trespassers on our private pathway. Eight-hundred miles and I had seen no one walking who was not coming to see me. This was again the case.

But these two were so far out of context and such an odd pair that my brain took a few ticks to process who they were. As they neared, though, the neurons in my brain made the necessary connections.

"Hey Ned!" Becky Baird called out.

Sean Willison accompanied her. Both possess fine sets of teeth they show often.

Becky and Sean

When I last spent time with Becky she powered a raft — in which rode my daughter Anna and two other little girls — through splashing whitewater on Birch Creek in Interior Alaska. Becky and her husband and daughter joined us and other friends for an 8-day trip during which we saw wolf tracks on every gravel bar we stopped at. I remembered her skill at the oars and, like Sean, her ever-present smile.

"You're almost done man!" Sean said.

"Nice work!" Becky said.

Both worked State of Alaska jobs in the oilfields; Sean had the same job Brian Jackson once had. We chatted a bit, and I told them I didn't know what accommodations awaited me at the end of the line.

"I feel stupid for not having made any arrangements. I figured it would have worked itself out by now."

"Well, it just did," Becky said. "You can sleep in the state workers' quarters with us and I'll make you dinner tonight."

Trail magic is a relatively common phenomenon along more traveled thru-hikes, such as the Appalachian and Pacific Crest trails. But when a trail has been thru-hiked by one person just twice, trail magic doesn't happen as often. A calm contentment warmed me as I watched Becky and Sean walk back over the wet tundra to their truck. My biggest immediate worry had vanished.

According to a little sign near a pipeline waterway, only 0.99 of a mile remained in our summer of hiking. We had run out of scenery; the curve of the Earth had hidden the Brooks

Range behind us. The arctic plain spread out around us in flat greens and yellows, with not a dry step to be had. Big gray sky spread like an ocean overhead. Every stride in that last mile took us closer to giant biscuit tanks and a few towers that looked like smokestacks with whitish flames bursting from the top (as natural gas was flared off).

The ending of every trip is a change in routine, and a big one was ahead for me. Finishing in the "dreadful place" George Schaller described would not be like setting up my tent next to John Arntz's on a dry aspen-and-black-spruce bench scented of sage.

But there was still the thrill of finishing something that I had started. Walking on, I spoke a silent mantra with every step:

Thank you, thank you, thank you, thank you.

I didn't really know to whom I was sending those thanks. It just seemed the right thing to do.

* * *

Soon, it was just ahead: Mile 0.

Cora and I crossed one more footbridge and took a few more steps on crunchy gravel.

SUMMER OF GRAVEL AND STEEL

And then it was over.

Sean and Becky had returned in their truck. Sean took a photo of me and Cora standing beneath the Mile 0 sign, where the pipeline from our perspective dove into the ground like an earthworm and disappeared.

Two Alyeska workers were also there at the finish. One knew of me and the hike. The other expressed surprise that someone had walked there from Valdez. They both shook my hand.

Becky and Sean gave Cora and me a ride in the truck to fulfill a hope of mine. They drove the network of gravel roads around Pump Station 1 until we reached the salt water of Prudhoe Bay.

Cora and I popped out of the truck onto the spongy wet tundra. I strode to the ocean and stood on the shore, gazing at the flat, gray infinite sea that continues to the top of the world. I knelt, dipped my fingers into the sea, and lifted them to my lips, tasting the salt water as I had done almost 100 days ago in Port Valdez.

I wanted to keep going. To find a boat and shove off into the Arctic Ocean, continuing the simple life of keeping fed and on course and seeing what was ahead and sleeping under the stars. But then I remembered that ocean would be covered with ice soon. And my daughter's 11th birthday was coming up in a few days and Kristen had planned a party in Fairbanks.

* * *

Back at the state workers' quarters, Becky made us a spaghetti dinner. Sean led me to a musty room with a desk and a bed, the first mattress I would sleep on since Jillian had set me up in Coldfoot weeks before.

In the foreign, overheated environment smelling of disinfectant, Cora wouldn't leave my side. She even came into the bathroom with me as I indulged in a hot shower for the first time since Coldfoot. I hadn't missed taking showers, but the warm water relaxed me.

Cora wasn't as calm, panting in the bathroom

I massaged her ears and tried to sooth us both.

"Yeah, this is weird, sweetheart. But soon you'll be back to your couch at home."

Back in the sleeping room with Cora, I enjoyed the strange luxury of a chair and desk while writing in my journal. There came the final surprise of the trip.

I looked up from my journal at the computer terminal. Brian Jackson looked back at me.

His photo was on a security badge his co-workers had left hanging there.

I sat at the desk he had used on trips to Prudhoe before he died. The computer in front of me once displayed draft picks of the Green Bay Packers as Brian kicked back after work. He spent his days monitoring oil spills for his job with the Alaska Department of Environmental Conservation.

I had thought about Brian when starting this trek in Valdez. His death helped motivate me to overcome my own inertia of normalcy. To get unstuck from my over-predictable life. It's always right to go. And here he was, showing up at the end to remind me that nothin' lasts forever.

I took Brian's badge and held it in my fingers.

"Dude, we executed!" I said. My eyes misted up.

I blinked, then turned to look out the window at the still-light August sky, thinking of how that light had bathed us all summer, turning my skin brown and making things easy to read without glasses.

The clouds out the window had started to spit snow. Another short summer of life was over.

25

Afterword

Right after completing the last steps of my first pipeline hike, I paused my day job, sat my butt in a chair every morning, typed on my Macintosh Classic II and busted out the manuscript for *Walking my Dog, Jane*. I had it to the publisher in less than a year.

This book was not like that. My daughter somehow went from 10 to 17 when I was tapping away on it in our little guest cabin.

When something takes that long, things change. Clutch Lounsbury, who was so nice to my cousin Heather as well as Kristen, Anna and me, passed away a few months before I finished this. Good man, he was.

Cora's best friend and one of my best friends Fluffy also left us. I still find her cream-colored wool on our walks.

My brother-in-law David Bartecchi died in the Rogue River while freeing other boaters. He was the ultimate yes man.

Somewhere during the period from hike to this book I also got a cellphone.

There are many people I didn't mention in the book who helped me.

Heather Best and Jeff Oatley gave me more than 100 Via coffee singles. Jeff also delivered buffalo burgers to me and Andy Sterns south of Delta Junction and camped with us for a night there.

Luke Boles showed up down there, too. He also got a kennel to Deadhorse on deadline so Cora could fly home.

Luke's wife Emily Youcha hid diet Cokes for me at strategic spots along the path. During a summer when I drank almost nothing carbonated, finding those was a big deal.

Susan Sharbaugh texted me Yankee scores every night on my inReach. I always looked forward to that blinking green light at 9 p.m.

Elizabeth Schafer found me and Cora as she was driving home to McCarthy. She walked with me during a sunny afternoon and laughed the Elizabeth Laugh.

Alyeska security guys and gals always stopped to chat, and gave me information about whether the next creek had a bridge and apples from their lunches.

Ed Plumb let us use his Alaska Range cabin for an overnight. That allowed us to finally photograph Ted (Wu), Ned, and Ed in a bed.

Ted, Ned and Ed

Judy Hicks of Delta Junction gave us a cool place to stay on the Tanana River and asked for nothing in return.

It warmed my heart to revisit my favorite homesteaders,

Jeanette and Robert Walker, on their beautiful spread north of Delta.

Sheri Tingey of Alpacka Rafts loaned me a Scout that got me across the Salcha and Chatanika Rivers.

Heather Kelly and Brad Hurd of Heather's Choice gave me a bunch of smoked salmon dinners and many other dehydrated goodies.

Bob and Judi Sterns transported Bob Gillis to me, fed me at Pump House, and created Andy Sterns.

Audrey Bohl fed and watered me and John Arntz at Five-Mile Camp north of the Yukon River.

Sarah Ast drove her sons Garrett, Brody and a Wiffleball set to the Arctic Circle to camp with me. She also brought her sister Kristen and niece Anna and helped extinguish the toilet-paper grass fire.

David Fee and Taryn Lopez hiked me out a food drop box and Andy Sterns just south of the Koyukuk River.

Jack Reakoff stuffed us with moose and potatoes during a visit to his home with Knut Kielland and Doug Best.

A trucker named Roger stopped his rig smack in the

middle of the Dalton Highway to hand a few muffins to Cora and me. He even turned off his engine.

Officials with the University of Alaska Fairbanks Institute of Arctic Biology who run Toolik Field Station let Cora and I stay there for a night as long as she was on a leash and didn't kill any ground squirrels. That was really skookum as they were full up for the summer with researchers.

Jason Clark, a UAF graduate student and also my neighbor, greeted us at Toolik and made things easy for us. He later solved an iPad problem while laying belly down on the tundra.

Ruby, a graduate student from Chicago who was at Toolik for the summer, gave Cora lots of pets and drove Eric Troyer and me back to the pipeline in a pickup.

Eric Troyer was not my favorite person for the 24 hours after I saw his 400 red suggestions to each of my chapter drafts. But I loved him two days later, and the many days beyond that.

An extra-special thank you to the rock-solid Michelle Charlton of Northern Alaska Tour Company. And all her drivers. Especially Tony, who got cheerily destroyed by mosquitoes while walking a box to me across a few acres of northern tundra.

Our friends the Carlsons — Audra, Chris, Ian, Ella and Nora — were the beautiful neighbors and friends they have always been. Chris' brother Paul also loaned his Summit Lake cabin when we really needed it.

And finally to my girls Kristen and Anna, who went along with a disruptive idea and made it more fun and enriching than I thought it would be.

Thank you, thank you, thank you, thank you.

26

About the author

Ned Rozell has pressed a few million footprints into Alaska. He has written that many words about it too. He is the author of 8 books — including the prequel of this one, *Walking my Dog, Jane* — and has written 2,000 weekly science/natural history stories about Alaska as part of his day job for the University of Alaska Geophysical Institute. See them at https://www.gi.alaska.edu/alaska-science-forum.

www.ingramcontent.com/pod-product-compliance
Lightning Source LLC
Chambersburg PA
CBHW070802040426
42333CB00061B/1785